The Industrial Labor Market and Economic Performance in Senegal

The Industrial Labor Market and Economic Performance in Senegal

A Study in Enterprise Ownership, Export Orientation, and Government Regulation

Katherine Terrell and Jan Svejnar

Westview Press
BOULDER, SAN FRANCISCO, & LONDON

This Westview softcover edition is printed on acid-free paper and bound in softcovers that carry the highest rating of the National Association of State Textbook Administrators, in consultation with the Association of American Publishers and the Book Manufacturers' Institute.

All rights reserved. No part of this publication may be reproduced or transmitted in any form or by any means, electronic or mechanical, including photocopy, recording, or any information storage and retrieval system, without permission in writing from the publisher.

Copyright © 1989 by Westview Press, Inc.

Published in 1989 in the United States of America by Westview Press, Inc., 5500 Central Avenue, Boulder, Colorado 80301, and in the United Kingdom by Westview Press, Inc., 13 Brunswick Centre, London WC1N 1AF, England

Library of Congress Cataloging-in-Publication Data
Terrell, Katherine D.
  The industrial labor market and economic performance in Senegal:
a study in enterprise ownership, export orientation, and government
regulation / by Katherine Terrell and Jan Svejnar.
    p. cm.
  Bibliography: p.
  Includes index.
  ISBN 0-8133-7801-X
  1. Labor market—Senegal. 2. Senegal—Industries. I.
Svejnar, Jan. II. Title.
HD5847.7.A6T47 1989
331.12′09663—dc20   89-8997
                                                                  CIP

Printed and bound in the United States of America

∞   The paper used in this publication meets the requirements of the American National Standard for Permanence of Paper for Printed Library Materials Z39.48-1984.

10   9   8   7   6   5   4   3   2   1

To our parents

Contents

Foreword, Guy Pfeffermann ix
Acknowledgments xi

1 Introduction 1

2 The Economic Performance of Senegal, 1960-1987 5

    2.1 Structure and Growth of the GDP, 5
    2.2 Structure and Growth of the Industrial Sector, 8

3 Labor Market Institutions and the Industrial Relations System 13

    3.1 Specific Institutional Aspects, 15
        3.1.1 Hiring, 15
        3.1.2 Employment Contracts, 16
        3.1.3 Layoffs, 17

    3.2 Wages, Bonuses, and Supplements, 17

4 Basic Demographic, Employment, and Wage Statistics 21

    4.1 The Demographic Situation, 21
    4.2 Labor Force, Employment, and Unemployment, 22
    4.3 Industrial Employment in the Modern Sector, 24
        4.3.1 Employment of Permanent Workers, 25
        4.3.2 Employment of Temporary Workers, 28
        4.3.3 National Composition of Employment, 31

    4.4 Earnings, 34
        4.4.1 Components of Total Compensation, 34
        4.4.2 Evolution of Wages by Skill Groups, 38
        4.4.3 Evolution of Total Compensation by National and Skill Groups, 40

5 Labor Market Regulations, Industrial Relations, and Enterprise Performance   43

   5.1 The Position of Government Officials, 43
   5.2 Managers' Views, 45
        5.2.1 Restrictive Labor Regulations, 45
        5.2.2 Recruitment and Training, 46

   5.3 Workers' Views of Unions, 49
   5.4 Trade Union Perspective, 50
   5.5 The Noncooperative Outcome, 51

6 The Productivity Effects of Labor Force Structure, Enterprise Ownership, and Foreign Trade Variables   53

   6.1 Overall Production Function Estimates, 56
   6.2 Industry-Specific Production Function Estimates, 69

7 Econometric Estimates of the Determinants of Wages   101

8 Industrial Wages and the Extended Family   107

   8.1 Size of Extended Families, 108
        8.1.1 Nuclear Families, 108
        8.1.2 Non-nuclear Families, 110

   8.2 Relationship Between Earnings and the Extended Family, 112
   8.3 Secondary Income, 114

9 Summary and Policy Implications   117

   9.1 Summary, 117
   9.2 Policy Implications, 119

References   123
Index   125

Foreword

Nearly 25 years ago, I interviewed industrial workers and their employers in Senegal. <u>Industrial Labor in the Republic of Senegal</u> (Praeger, New York, 1968) tells the story. Labor market imperfections, mostly inherited from French colonial legislation and institutions but also rooted in the extended family system, were so egregious as to stifle development. The replacement of expatriates by African workers was proceeding at a snail's pace, government and union wage policies were curbing employment growth, vocational schools were turning out graduates unsuitable for industrial employment. Wage and other regulations intended to advance the workers' welfare failed to achieve their objectives. The extended family system obliged workers to support a multitude of unemployed hangers-on, all the more numerous owing to polygamism, and this undercut incentives to strive for better pay and to save more. Labor market distortions encouraged enterprises to use capital-intensive methods of production rather than to expand employment. And a parallel "informal sector" had sprung up of firms small enough to evade labor legislation.

Like the Three Musketeers, Terrell and Svejnar have revisited the Senegalese labor market 20 years or so later. Undoubtedly, some things have changed. Fewer expatriates are now working in Senegal. But the striking picture which emerges is that things mostly have remained unchanged. Labor regulations are still pervasive and largely self-defeating. Labor productivity remains low. The supply of skilled workers remains limited. And the burden of the extended family on wage-earners has become even greater. The authors document this unhappy situation very carefully with the help of interviews and detailed surveys and ask themselves to what extent Senegal's uncommonly poor growth performance can be traced to labor market imperfections. They draw policy implications and make recommendations that address some of the root causes of stagnation. These should be of great interest to the government, which has been striving to improve the structure of the economy so as to accelerate growth in a sustainable manner. The book will also be of interest to other developing countries whose labor market institutions have much in common with Senegal's.

Guy Pfeffermann
Economic Adviser to the International Finance Corporation
World Bank Group

Acknowledgments

The research for this monograph was in large part funded by grant no. 674-03 of the World Bank Research Committee. We are especially indebted to Guy Pfeffermann for conceiving the idea and assisting us in launching and implementing the project. Without his guidance and supervision the study would not have been realized. We are also grateful to a number of other individuals and organizations that provided invaluable help during the research project: Mark Leiserson contributed enormously to the project's formation, and the World Bank Senegal Division, the Industrial Strategy and Policy Division, and the World Bank Resident Mission in Dakar assisted in our fieldwork. We are particularly grateful to Kemal Dervis, Christian Duvigneau, Nicolas Gorjestani, Frans Kaps, Christopher Redfern, and Eugene Scanteie.

We extend special thanks to Mme. Thiongane, the Director of Statistics at the Ministry of Finance in Senegal, for providing us with the data; to Prospere Youm, the Director of Forecasting at the Ministry of Finance, for providing data on the effective rate of protection; and to Alhousseney Sy, the Chief of the Census Bureau, for collaborating with us on the sample survey. The trade union representatives and numerous enterprise managers who provided us with their perspectives have made valuable contributions to our understanding of Senegal's industrial sector and its labor market institutions.

Finally, we thank Bela Balassa, Guy Standing, Gerald Starr, Erik Thorbecke, and the participants at the World Bank seminar for their many helpful comments, Sathi Pandian and Seng-Lee Wong for their excellent computational assistance, and Lauree Graham for invaluable help in the preparation of the manuscript. All remaining errors are our own.

Katherine Terrell
Jan Svejnar

1

Introduction

Senegal is in the grip of a long-term economic crisis. Since 1960, it has experienced the smallest growth in gross domestic product (GDP)--2.3 percent--of any African country not affected by war or civil strife. Moreover, data suggest that the GDP per capita has been at best stagnant and probably falling over the last twenty years as the productivity and real income of most of the population have not been raised since independence. Since the late 1970s the Senegalese government has recognized the need to embark on a sustained and long-term adjustment program and has searched for appropriate policies to pursue this goal. In this context, an important objective of this study is to provide an up-to-date analysis of several issues that are pertinent to the ongoing debate in the choice of appropriate adjustment policies in Senegal. Since many problems examined in this study have been encountered in other countries, we hope that our analysis also has broader policy ramifications.

One of the structural causes of the current economic crisis is believed to be the inability of the modern sector to raise its overall share of employment as well as its productivity and average real earnings in the past two decades. The sluggish performance of the modern (formal) industrial sector has been the subject of major debate, given that the sector had a relatively mature base at independence, has received considerable investment since then, and has enjoyed significant protection as well as preferential trade agreements in both the European (EEC) and West African (CEAO) Economic Community.

Most economists agree that employment in the informal urban sector has grown over the past two decades but that the formal sector has not generated much employment and has instead switched to more capital-intensive methods of production. It is also suggested that public enterprises have been the only significant providers of jobs in the modern sector. A recent study by Mingat (1982) claims, however, that employment generation in the informal sector has slowed. What has increased, according to this study, is the level of open unemployment in the urban labor market. These opposing views illustrate the prevailing lack of basic knowledge about the structural features and functioning of Senegal's industrial sector and labor market. They have also provided us with an important motivation for undertaking the present study.

Our goal is to examine Senegal's industrial labor market and relate its principal features, together with the structure of enterprise ownership and government policies, to the country's economic performance. Our primary aim is to assess how the employment and wage setting system affects allocative and productive efficiency of industrial enterprises. Because efficiency is widely believed to be tied to the structure of ownership and to the government's industrial and trade policies, our secondary aim is to estimate the total factor productivity effects of different forms of enterprise ownership, the degree of effective protection of firms, and enterprise export orientation.

The emphasis of our study is on the modern industrial sector, although links with other sectors in the economy are also explored. The modern industrial sector includes large as well as relatively small firms (e.g. ones with 10 employees) and for the purposes of this study it is defined as that part of industrial activity which is systematically recorded by official statistics. Our focus on the modern sector is motivated by the fact that (1) its economic performance is believed to have been largely impeded by government regulations and could therefore be expected to improve significantly as a result of more appropriate policies; (2) it provides the majority of Senegal's exports; and (3) in the long term it should play a leading part in spearheading economic growth, in generating employment, and in absorbing and diffusing new technologies. Moreover, relatively reliable data and qualitative information on this sector were available or could be gathered within a relatively short period of time. Finally, knowledge of the characteristics and performance of the formal labor market provides an important yardstick against which one can compare and evaluate the performance of the informal labor market.

The main part of our analysis relates to the period from 1980 to 1986. The selection of this time period was guided by both the need for policy relevance and data availability. However, to provide a fuller understanding of the main issues and the underlying dynamics, the findings from this period are compared with those of Pfeffermann (1968) for the 1960s and, whenever possible, with those of Svejnar (1984) for the mid 1970s.

As mentioned above, our main aim is to provide an understanding of the functioning of the industrial labor market and evaluate the effects of the employment and wage setting practices on enterprise efficiency. In undertaking this task, we pay particular attention to government wage and employment regulations; the role of trade unions and the organizational characteristics of firms; the national and skill composition of the labor force and the corresponding labor productivity and wage differentials; the existence of training schemes and skill bottlenecks; the effect of education, experience, and training on worker earnings; and the relationship between a worker's income and the migration of family members from the countryside.

Our approach is to combine institutional analysis with econometric estimation. In Chapter 2 we provide a brief survey of Senegal's economic performance, focusing on the evolution of the industrial sector. In Chapter 3 we describe the main features of the Senegalese industrial relations system, and in Chapter 4 we discuss basic statistical evidence relating to the characteristics of the labor force, employment, and worker earnings. Chapter 5 contains an institutional analysis of the impact of labor regulations and the industrial relations system on economic performance in industry; chapters 6 and 7 present the corresponding econometric analysis. Chapter 6 also contains estimates of the productivity effects of the national and skill composition of the labor force, different forms of enterprise ownership, and foreign trade variables. This analysis is based on data from a relatively sizable enterprise sample which we collected during the course of the research project. Chapter 7 presents econometric estimates of the determinants of worker earnings in a smaller (sub) sample of modern sector industrial enterprises. Chapter 8 is devoted to exploring the relationship between industrial wages and the size of the extended family. Finally, Chapter 9 presents the summary and policy conclusions.

2

The Economic Performance of Senegal, 1960 - 1987

At independence, Senegalese nationals inherited a relatively well-equipped physical and social infrastructure and a GDP per capita that was larger than that of either the Ivory Coast or Cameroon. Over the quarter of a century that followed, however, the economy failed to raise either labor productivity or real per capita income. Official data actually suggest that real GDP per capita decreased by 4 percent between 1967 and 1987,[1] while real output per employee and real earnings per employee in manufacturing dropped 31 percent and 33 percent, respectively, during the 1974 - 1985 period.[2] As a result, by 1983 Senegal's per capita income of $440 (using the 1987 exchange rate) was barely half that of Cameroon and the Ivory Coast, placing it near the bottom of the ranks of lower middle-income oil importers. The following two sections describe how the structure and growth of gross domestic product changed between 1960 and 1986.

2.1 Structure and Growth of the GDP

Despite certain structural changes, the Senegal economy strongly reflects the production patterns established during the colonial period. As can be seen in Table 2.1, Senegal already had a relatively large secondary and tertiary sector at independence

---

1. This result is based on calculations from the following data: current GDP in 1967 = 205.5 billion CFA (African Financial Community) Francs, current GDP in 1987 = 1419 billion CFA Francs, GDP deflator in 1967 = 42, GDP deflator in 1987 = 177.6, population in 1967 = 4,117,000, and population in 1987 = 6,969,000. The official data are taken from World Bank (1988 - 89), volume 2, pp. 486-487.

2. See ibid.

in 1960.[3] In the past twenty-seven years, the share of production has increased by about 50 percent in the secondary sector and fallen slightly in the primary and tertiary sectors.

Essentially, the traditional part of agriculture still produces millet and cattle for domestic consumption, and the export-oriented part produces groundnuts. Domestic agricultural production does not cover consumption--nearly one-half of Senegal's basic food stuffs are imported.[4] As we show later, the modern sector produces primarily light industrial goods and is highly concentrated regionally and in terms of ownership.

TABLE 2.1
Senegal: Structure of Production

Sector	Share of GDP (%) (in current prices)				
	1960[a]	1970[a]	1980[a]	1986[b]	1987[b]
Primary	24.3	24.1	19.1	21.8	21.6
Secondary	18.6	21.5	24.9	24.8	25.7
Tertiary	57.1	54.4	56.0	53.4	52.7

Source: [a]World Bank, Senegal: Country Economic Memorandum, 1984, p.4.
[b]International Monetary Fund (IMF) Senegal - Statistical Annex, 1988, p. iv. (Figures for 1987 are preliminary).

The limited change in the structure of production reflects the general lack of dynamism in the economy. As can be seen in Table 2.2, real GDP growth declined from an average of 2.5 percent a year throughout the 1960s to an average of 1.9 percent

---

3. The primary sector contains agriculture, livestock, fishing and forestry. The secondary sector contains industry and mining, construction and public works, and energy. The tertiary sector is composed of transportation, commerce, and government and other private services.

4. For a recent account of issues related to Senegal's agriculture, see Gersovitz and Waterbury (1987).

in the 1970s, but it accelerated to over 3 percent in the 1980s. With population growth estimated at 2.8 percent throughout these two and a half decades, the trend in the per capita GDP has been at best stagnant and probably declining.

TABLE 2.2
Senegal: Estimated Real Annual Growth in GDP,
Total and by Sector

Sector	1960-70[a]	1970-80[a]	1980-86[a]	1987[b]
Primary	3.0	-0.4	2.7	2.8
Secondary	4.2	2.7	4.2	7.8
Tertiary	1.7	1.3	2.2	2.7
Total	2.5	1.9	3.1	4.0

Note: Based on series in 1979 constant prices.
Source: [a]World Bank, Senegal, Country Economic Memorandum, 1984, p. 4; World Bank, Senegal, An Economy Under Adjustment, 1987.
[b]IMF, Senegal - Statistical Annex, 1988, estimates.

Most of the decline in the real growth of GDP came in the second half of the 1970s. As a result of severe droughts in 1977-1978 and declining production incentives, net output in the primary sector declined by an average of 3.9 percent per year from 1975 to 1980. Growth of output in the secondary sector also slowed down to 2.2 percent per annum during the latter half of the 1970s, down from an average growth of 3.2 percent per annum in the first half of the 1970s and 4.2 percent per year during the 1960s. Only the service sector (which includes public administration) grew at a faster rate in 1975-1980 than in 1970-1975.

The growth of the public (government) sector in 1975-80 has been steady, with the share of government consumption in final demand rising from 14.9 percent in 1970 to 19.8 percent in 1983. A particularly rapid increase occurred between 1975 and 1980, when the share jumped from 15.2 to 22.4 percent. The growth of public expenditures during this period is explained partially by the large costly programs that the government implemented in these years in order to diversify the economy and develop Senegal's human resources. Some economists maintain that during this period an excessive share of public investment was used for social purposes, to maintain consumption patterns of the urban elite, or to extend public control of the economy rather than to create new productive capacity.

Although economic mismanagement, together with poor rainfall and adverse external factors, had already eroded the country's strength, an acute drought in 1977-1978, the 1979 increase in oil prices, and two additional poor groundnut crops in 1979 and 1980 almost crippled Senegal's economy. In 1980, real GDP fell by 1.5 percent with a further 2.4 percent decline occurring in 1981. The deficit on the external current account reached US$584 million, or 23.6 percent of GDP in 1981.

It is argued that the roots of the liquidity crisis in 1980-1981 lay in the failure of the government to control the growth of total expenditures. In the late 1970s the government began to recognize the shortcomings of its development plans and started efforts to adjust to the crisis.

In 1982 and 1983 there was a strong recovery in GDP, with estimated growth rates of 15.1 percent and 2.7 percent, respectively; however, following the severe drought of 1983-1984, the whole economy suffered a recession of the same amplitude as that of 1980-1981. Even so, because of the two good years before the drought and a 4.1 percent annual growth in real GDP during the 1985-1987 period, there has been overall resurgence in the economic activity in recent years. (See Table 2.2.)

2.2 Structure and Growth of the Industrial Sector

As seen above, the principal engine of growth in Senegal's economy from 1960 to 1986 was the secondary sector, with industry accounting for over 60 percent of its value added. However, as Table 2.3 demonstrates, since 1976 even the growth in the industrial sector has been disappointing. Since this sector is a crucial element of Senegal's economy, we next discuss its composition and growth patterns.

The industrial sector is essentially made up of three resource-based export-oriented industries--groundnut processing, phosphate mining, and fish processing--and a range of import-substituting light industries. It is a highly concentrated sector: 140 of the 600 modern industrial enterprises account for over 94 percent of total sales. Virtually all subbranches of industry are dominated by 1 or 2 companies, usually either government controlled or foreign owned. About one-half of all enterprises are under French-majority ownership. The government holds majority interests in 17 of the most important industrial companies. Hence foreign and Senegalese government ownership are notable and interrelated features of the industrial sector with the former being in many respects more significant than the latter.

In general, the level of production in the industrial sector between 1977 and 1981 was at or below the 1976 level. Growth resumed in the 1982-1985 period but a decline occurred in 1986. (See Table 2.3.) The textiles, clothing, and leather sectors, the wood products sectors, and the paper producing sectors have reported the strongest performance over this period, although, as we will show in Chapter 4, this performance has not been fully reflected in employment growth. Moreover, some sectors, notably chemicals, construction materials, and machinery and equipment, have suffered a serious decline in production since 1977.

Manufacturing has been heavily oriented toward the domestic market, with exports representing less than 20 percent of total industrial sector sales. Processed primary products go to Europe (especially France), and light manufactured goods are traded almost exclusively to the CEAO trade zone neighbors. Despite preferential tariff arrangements in both the EEC and CEAO markets, export growth and diversification has been slow. In 1986, processed primary products (groundnuts, phosphates, petroleum products, and fish) still constituted two-thirds of Senegal's total exports of goods and 90 percent of the country's exports to developed country markets. Excluding groundnuts, exports declined by almost 20 percent in real terms between 1977 and 1984, although some recovery has occurred since 1980. Only fish exports demonstrated consistent real growth over the period, but further growth is expected to be limited due to the exhaustion of this valuable resource.

The trade regime in Senegal is considered to have been a prime example of overlapping policies with contradictory effects. Until very recently, nominal tariffs were high (ranging from 25 percent

TABLE 2.3
Senegal: Indices of Industrial Production
(1976=100)

	1978[a]	1981[a]	1983[a]	1985[b]	1986[b]
Mining	95.9	107.4	84.3	112.6	107.7
Food Production	91.6	88.5	109.7	103.2	99.6
Textiles, Clothing and Leather	101.0	110.4	139.9	133.9	123.0
Wood Products	111.0	127.1	133.0	119.0	100.2
Paper	113.6	120.6	137.5	120.1	130.5
Chemicals	104.2	85.5	73.1	77.9	51.9
Construction materials	103.4	94.7	128.6	94.1	83.8
Machinery and Equipment	110.6	61.8	93.6	78.1	41.0
Electricity and Water	121.5	134.6	148.7	155.0	166.4
General Index without oil processing	106.8	108.2	111.4	119.6	112.5
General Index	98.2	97.2	107.1	107.6	99.6

Source: [a]Direction de la Statistique, Indice de la Production Industrielle, in World Bank, Senegal, An Economy Under Adjustment, 1987, p. 96.
[b]IMF, Senegal - Statistical Annex, 1988.

to 90 percent) and escalated sharply for finished goods. Quantitative import restrictions existed for about 160 commodities, with significant effects on domestic market prices. The average level of protection in the domestic market was uneven and high, probably equal to at least 50 percent of the cost, insurance, and freight (CIF) value of competing world products (World Bank, 1984, p. 69). The gradual lowering of tariffs and diminution of quantitative restrictions since the mid 1980s has led to a much greater opening of Senegal's economy. As our estimates in Chapter 6 indicate, one can expect this new industrial policy to result in a significant increase in efficiency of the surviving firms.

In summary, Senegal's economy has performed poorly throughout the 1970s and early 1980s, with growth resuming at a somewhat faster pace from the mid 1980s on. As for the industrial sector, the rate of growth of output was above average until the mid 1970s and, after a hiatus of several years, again in the 1980s. However, several industrial sectors, including chemicals, construction materials, and machinery and equipment (mechanical industry), have experienced decline since the mid 1970s. In the chapters that follow, we examine government policies and other factors in the labor market that may have contributed to this unimpressive performance.

3

Labor Market Institutions and the Industrial Relations System

In this chapter we describe the principal features of the industrial labor market and the industrial relations system. The modern and traditional sectors differ widely in terms of worker earnings and the governmental regulations concerning hiring, layoffs, employment contracts and working conditions. The modern sector is, to a large extent, unionized, and it is subject to a relatively rigid system of regulations; the traditional sector is informal, operating on the basis of market forces and traditions.

The modern-sector labor market institutions are in many respects similar to those of other French-speaking African countries. The trade union movement is weak and closely linked to the government.[1] Industrial relations are to a large degree governed by Senegal's 1961 Labor Code and several subsequent legal provisions.[2] The 1961 code closely resembles the 1952 French Overseas Labor Code and grants considerable authority to the government and its agencies. The code greatly limits employers' discretion over hiring, layoffs, form of employment contracts, and working conditions. In the case of trade unions, it imposes permissible bounds on their activities. The code also provides general guidelines for the process of wage determination.

The code is supplemented by the National Collective Agreement (NCA) of May 27, 1982, which was signed by the main trade union federation--Confederation Nationale des Travailleurs du Senegal (CNTS)--the principal employer associations at the time--Union Syndicale des Industriels (UNISYNDI), Syndicat des Commercants, Importateurs et Exportateurs de l'Ouest Africain

---

1. The weakness of unions derives inter alia from their fragmentation, poverty, inefficiency of operation, corruption of middle-level officials, and the gulf between the leaders and the workers. See Pfeffermann (1968, ch. 5) and Martens (1982).

2. The references in the text are to the May 1984 edition of the Labor Code.

(SCIMPEX), and Sydnicat des Patrons de l'Ouest Africain (SYPAOA). The NCA explains in greater detail the system of industrial relations outlined by the code.

The three main actors on the labor scene are the government (Ministry of Public Affairs and Labor), the CNTS, and the employers' associations. These three "social partners" form the so-called Tripartite Commission, which determines (annually since 1981) the rate of increase in the base wages (salaire de base) of various professional (skill) categories. The crucial variable set by the Tripartite Commission is the minimum industrial wage, *salaire minimum interprofessionel garanti* (SMIG). It constitutes the benchmark wage to which the wages of higher-skilled workers and employees are loosely related. Under usual practice, the Tripartite Commission determines the new level of the SMIG and a percentage increase in the basic salary for each skill group (e.g., 7 percent for the most skilled workers, 15 percent for the least skilled workers and intermediate percentage increases for intermediate skill categories). The terms of the accords of the Tripartite Commission are extended to all other sectors through the Extension Decree.

An employee's base wage (salaire de base) can be augmented by various salary supplements (sursalaires) and bonuses (primes), ranging from seniority and performance bonuses to benefits for transportation and difficulty of working conditions (compensating differentials). The supplements and bonuses are in principle negotiated between trade unions and management, but in practice the supplements are frequently determined by the employer, and many bonuses are unilaterally predetermined by the Tripartite Commission, NCA, or the government. The supplements and bonuses constitute a sizable part of total compensation (about 30 percent for unskilled workers and up to 80 percent for professionals and managers) and they hence have a major impact on the level of total earnings and the labor cost.

Despite the fact that worker earnings in Senegal are relatively high by international standards, industrial relations continue to be adversarial rather than cooperative in nature. However, strike activity is limited somewhat by the government regulation of industrial relations which includes compulsory arbitration in case of impasse. Historically, the trade unions have de jure operated as independent organizations.[3] However, in practice they are closely linked to the government, and their views (e.g., on labor market regulations) are reflected in the laws, decrees and official declarations.

A major aspect of Senegal's collective bargaining is hence the tripartite negotiation at the "highest levels" of the government. The work of the Tripartite Commission is heavily influenced by these "high-level" discussions. The next important stage of bargaining occurs at the firm (plant) level where union stewards (delegues de personnel) take up worker grievances and negotiate some fringe benefits. Interestingly enough, the worker is frequently expected to share the grievance compensation with the steward who helped win the case.

3.1 Specific Institutional Aspects

This section describes government regulations in four important areas: hiring, employment contracts, layoffs, and wage setting. These regulations impose the most binding constraints on the functioning of the labor market and are likely to have major efficiency and welfare implications.

3.1.1 Hiring

Until late 1987 the process of hiring was regulated by Articles 195-199 of the Labor Code. These articles gave the government monopoly control over the hiring of workers by enterprises. The Labor Office (Service de la Main-D'Oeuvre) of the State Secretariat for Employment was in charge of exercising this legal control through its central and regional bureaus.

Article 199 of the code declared the Labor Office to be the only intermediary between the job seekers and potential employers and defined it as the referral service for workers of all categories. All job seekers were thus expected to register with the Labor Office, and employers were legally obliged to notify the office of all vacancies in their firms. Hiring of a worker was legally subject to a prior approval by the office. The existence of private employment agencies or direct hiring by employers was, until very recently, forbidden.

---

3. For a detailed institutional analysis of Senegal's industrial relations in the 1950s, 1960s and 1970s, see Martens (1982).

In practice, the employer would notify the Labor Office of vacancies, and the office would supply several (usually three or four) candidates that were prescreened on the basis of information supplied by the job seekers and the firm. The employer could keep requesting additional candidates until a suitable person was found for the job. The entire process could reportedly be rather time-consuming.

In 1987 a bill to eliminate the Labor Office's monopoly over hiring, to permit direct matching between workers and employers, and to establish private employment agencies was introduced in the National Assembly. The law was passed in 1988; as a result, direct worker-employer matching and referrals by private employment agencies are reportedly becoming more frequent.

3.1.2 Employment Contracts

The Labor Code (primarily Articles 34-38) permits three types of employment contracts: fixed-term, seasonal, and open-ended ("permanent"). The fixed-term contract is designed for the execution of specific tasks, and it legally can last as little as one day or as long as two years. In order to encourage long-term employment, the government has historically tried to limit the employers' use of this type of contract. In fact, Article 35 of the Labor Code prohibits more than two fixed-term contracts between a given firm and worker. A violation of this rule automatically grants an open-ended contract to the worker. Seasonal contracts can be made repeatedly by employers whose production contains a significant seasonal component. However, abuse of the seasonal contract also results in its automatic conversion into an open-ended contract.

Estimates by the Ministry of Industry suggest that only about 10 percent of employees in the manufacturing sector work under seasonal contracts; the fraction of workers with fixed contracts is also reported to be very low. The official estimates hence suggest that the application of the Labor Code has succeeded in installing the open-ended ("permanent") contract as the principal form of employment. However, as we show in Chapter 4, although this official picture is reflected in the ratio of the temporary workers' wage bill to the total wage bill, it underestimates the number of temporary workers that are used in any given year. The number of temporary workers amounts to almost 40 percent of the permanent workers.

As part of its attempt to reduce the rigidities of the modern-sector labor market, the government in 1977 approved a bill that, under certain conditions, allows employers to exceed the limitation on fixed-term contracts. The impact of this law is not yet clear.

3.1.3 Layoffs

An employer's ability to vary the size of the labor force through layoffs is regulated primarily by Articles 45 to 47 of the Labor Code. In particular, economic (as opposed to disciplinary) layoffs of workers with open-ended contracts must be approved by a labor inspector on the basis of a well-prepared file submitted by the employer. The inspector examines the evidence to determine whether the economic difficulties experienced by the enterprise in question are serious enough to warrant the requested layoffs. Article 47 also stipulates that internal reorganization of the firm may constitute a valid reason for layoffs, but in practice employers find this to be a difficult ground for justifying layoffs.

The labor inspector has up to 45 days to approve or deny the proposed layoff. This ruling can be appealed by either party to the Labor Directorate of the Ministry of Public Affairs and Labor. The decision of the Labor Directorate can in principle be appealed to the Supreme Court; in practice, however, it usually represents the final verdict.

Layoffs of workers with fixed-term contracts are possible only if (1) the employer can prove a serious offense on the part of the worker or (2) the employer can show that the performance of the enterprise has been seriously hampered by an "act of nature" (e.g., fire). Economic hardship brought about by a recession is hence not a valid reason for laying off workers with fixed-term contracts.

3.2 Wages, Bonuses, and Supplements

As we mentioned earlier, Senegal's wage setting system represents a mixture of tripartite and bilateral negotiations, as well as unilateral actions on the part of employers. The increases in the SMIG and in the basic salary of each occupational (skill) category have been determined annually since 1981 by the Tripartite Commission. The increase in supplements and bonuses is determined in part by the government, in part by negotiations between trade unions and management, and in part by management alone. In the case of an impasse in the management-union bargaining, the settlement is imposed by a government arbitrator. The compulsory arbitration system effectively limits, but does not eliminate, strike (lock-out) activity.[4]

As mentioned earlier, total earnings are composed of a base wage (salaire de base) and bonuses and supplements (primes and sursalaires). The bonuses and supplements often represent a sizable portion of total compensation, especially of the more educated employees. In fact the ratio of the basic salary to total compensation decreases rapidly with a person's education and professional qualification. With rare exceptions, worker earnings are linked neither to individual or group productivity nor to enterprise profitability. The four main bonuses received by workers in addition to their basic salary are:

(1) <u>Seniority Bonus</u>. This bonus is determined by Article 45 of the NCA and is based on the worker's length of service with a given enterprise. The amount is equal to 2 percent of the worker's basic salary after two years of service, and it increases by 1 percent each year up to twenty-five years.

(2) <u>Overtime and Night Work Bonus</u>. This bonus is accorded by Article 44 of the NCA for (a) night work, (b) continuous work of 10 hours, and (c) overtime work of 3 or more hours. The amount is equal to 3 times the hourly SMIG per day.

(3) <u>Transportation Bonus</u>. This bonus is determined by a mixed employer-worker commission on the basis of Article 46 of the NCA. In 1987-1988 it stood at 8,885 francs CFA (African Financial Community) per month (the cost of 95 trips on the first three sections of public bus transport).

(4) <u>Bonus for Dirty Working Conditions</u>. This bonus is based on previous NCAs and it varies across industries and professions. It is provided to workers who do not automatically receive workclothes from their employers. In the mechanical engineering sector the average bonus in 1987 stood at 4.675 francs CFA per month.

---

4. For a discussion of the nature and extent of strike activity see Martens (1982).

Numerous other bonuses exist. They raise significantly the basic salary, but very few of them are visibly linked to productivity or other performance indicators.

4

Basic Demographic, Employment, and Wage Statistics

Basic statistics about population, structure of employment, and wage level are important for understanding Senegal's labor markets. Moreover, they are also useful as background to the analytical work that follows in Chapters 5-7.

4.1 The Demographic Situation

Demographic data for Senegal are derived primarily from the 1960-1961 and 1970-1971 demographic surveys, the national census of 1976, the fecundity survey of 1978, and the labor migration survey of 1979. The data are of varied quality, and the statistics presented below should be used with caution. (In 1987 Senegal undertook another national census, the results of which should be available in 1990.)

According to the data, Senegal's population increased from 4,589 million in 1972 to 6,038 million in 1982. The annual rate of increase was a relatively high 2.9 percent during the 1972-1982 period, with the result that 53 percent of the population was less than 20 years old in 1982. In 1982 the life expectancy at birth was 48 years (47 for men and 49 for women), and the average number of surviving children per family was estimated at 7.16--a large number resulting in part from a high incidence of polygamy. National statistics indicate that in 1982, 32 percent of married men were polygamous, while 48 percent of married women were members of polygamous households. The incidence of polygamy may have actually risen over the past twenty years (see Chapter 7).

As we indicated in Chapter 2, most of the modern industrial activity is concentrated around Dakar. As a result, it is not surprising to find that 1,341 million people (22.2 percent of the population) reside in the Cap-Vert region (which contains Dakar) and that the population density in Cap-Vert is 75 times the national average.

4.2 Labor Force, Employment, and Unemployment

The data on overall labor force participation, employment, and unemployment are perhaps the least reliable of the demographic statistics. We report the official data for the sake of completeness, but we prefer to place more emphasis on the data from our enterprise and worker samples, which we present in the following sections of this chapter.

As of 1982, the number of people 10 years of age or older was estimated to be 4,111 million, of whom 2,274 million (55 percent) were less than 30 years old. The labor force in 1982 was estimated at 2,917 million, of whom 359,000 (12.3 percent) were classified as unemployed (where unemployed is defined as persons not gainfully employed and looking for work).

Over one-half of the employed population works in the traditional rural sector. The breakdown of employment by sectors, as reported by the Ministry of Planning for 1982 is:

Modern sector	905,000	(35 %)
Urban Informal sector	282,000	(11 %)
Traditional sector	1,371,000	(54 %)
Total	2,558,000	(100 %)

Within the modern sector, 317,000 of those employed are reported to hold permanent jobs, 494,000 hold seasonal jobs, 15,000 are classified as daily workers, and 79,000 work on a temporary or occasional basis. The striking aspect of these statistics is the large number of nonpermanent (especially seasonal) workers relative to the permanent ones. This finding runs counter to many widely held beliefs, and it is corroborated by data from our industrial sample. In contrast, within the informal sector a surprisingly large number of workers (223,000 out of 282,000) is classified as permanent. Given the nature of this sector, the definition of a permanent worker is presumably different -- probably meaning any worker who engaged in any informal sector activity for an extended period of time.

The official breakdown of total employment in 1982 by agriculture and resource extraction, industry, and services suggests that a sizable portion of employment in the modern sector (as classified above) is nonindustrial. In particular, the primary sector is reported to account for 74.3 percent of total employment, whereas the secondary and tertiary sectors account for 5.1 percent and 20.6 percent, respectively. Total modern sector employment therefore appears to be almost seven times the employment in industry (35 percent of the total vs. 5.1 percent of the total). A notable feature is the large size of the public sector (15.5 percent in 1980).

The unemployment problem appears to be most serious in the predominantly urban area of Cap-Vert. The 1979 labor migration survey, for example, reported that the unemployment rate in Cap-Vert was 19.4 percent; the national average stood at 6.8 percent in that year. Moreover, Cap-Vert was the only region in which the unemployment rate exceeded the national average, according to the official data.

TABLE 4.1
Senegal: Urban Labor Force, 1960 - 1980

	Number of Workers (in thousands)				Percentage of labor force	
	1960	1970	1976	1980	1960	1980
Modern Sector	88.3	122.4	158.7	195.5	35.0	34.3
Productive Sector	42.2	60.6	86.9	107.1	16.7	18.8
Agricul. & Fishing	1.5	4.9	5.7	8.5	0.6	1.5
Manufacturing	11.6	15.0	20.8	23.6	4.6	4.1
Public Sector*	46.1	61.8	71.7	88.4	18.3	15.5
Informal Sector**	147.0	205.2	241.1	275.7	58.2	48.4
Openly Unemployed	17.2	41.1	69.8	98.9	6.8	17.3
Urban Labor Force	252.5	368.8	469.6	570.1	100.0	100.0

* Civil Service and Non-Commercial Public Enterprises.
** Presumably calculated as a residual. The source does not identify methodology.
Source: Statistical Directorate, Ministry of Finance, Senegal.

The urban unemployment problem is also underscored in Alain Mingat's 1982 study. Table 4.1, which is assembled from Mingat's study, indicates that urban employment and unemployment both increased between 1960 and 1980, but that the level of unemployment increased much faster than did the level of employment. The study also suggests that formal sector employment increased faster than did employment in the informal sector. This finding challenges the traditional beliefs of most outside economic observers, but it is not clear what Mingat's source for informal sector data is. Bloch (1985) also figures and attributes them to the Statistical Directorate of the Ministry of Finance. He presumes that employment in the informal sector is calculated as a residual because no methodology could be identified. Our attempts to gather basic data on the informal sector lead us to conclude that the informal sector is one of the least understood activities in Senegal's economy and that data on this sector must be treated with extreme caution. It is for instance likely that the unemployed are efffectively part of the informal sector, although statistical information on this aspect of unemployment is scant.

4.3 Industrial Employment in the Modern Sector

As Table 4.2 indicates, the number of modern sector industrial firms appears to have increased sharply from 1963 to 1977, but has remained about the same since then. However, it is not clear whether the definition of the modern sector remained the same from 1963 to 1977; if not, the major increase in the number of firms during that period may be somewhat spurious. The definition during the past ten years has, in contrast, been uniform. Under this definition, a firm is considered to be part of the modern sector if it can be reached by the tax authorities. The important point is that even relatively small firms (e.g., with a dozen workers) are included. The data from 1977 to 1984 are hence fairly reliable. Further, they indicate that the much-discussed exodus of French entrepreneurs has not resulted in a significant decrease in the number of industrial enterprises. Our inquiries in this area suggest that most departing expatriates have been selling their firms to other expatriates or to Senegalese nationals. By the same token, it should be noted that there has not been any growth in the total number of modern sector firms.

TABLE 4.2

Senegal: Number of Modern-Sector Industrial Firms

	1963	1977	1980	1984
Food	30	120	106	102
Groundnut Processing	7	6	6	2
Textile and Footwear	19	18	23	26
Woodware and Packaging	7	21	14	16
Chemicals	7	28	36	38
Metal Engineering	22	30	35	44
Construction Materials	6	7	3	6
TOTAL	98	230	223	234

Source: Pfeffermann (1968, p. 4) for 1963. Direction de la Statistique for 1977, 1980 and 1984.

4.3.1 Employment of Permanent Workers

Table 4.3 provides the official data on employment of permanent workers by industry since 1977. Total manufacturing employment remained roughly constant between 1977 and 1982, (nearly 27,000), but it increased by about 2,000 jobs during the slight economic recovery of 1984. Generally, the modern industrial sector has not been a major creator of permanent jobs, although the overall stagnation disguises any variation across individual industries. In particular, textiles, clothing, and leather, wood, paper, and construction are among the industries whose level of permanent employment has decreased, whereas printing, chemicals, and mechanical industry substantially increased their employment of permanent workers since 1977. This finding is curious given that production in the former group (excluding construction) grew more rapidly than the average during this period while production in chemicals and mechanical industry declined. (See Table 2.3.)

Before attempting to explain this striking phenomenon, we present similar data for a sample of 80 industrial enterprises. The sample, which we designed jointly with officials in Senegal's

Ministry of Finance, accounts for about 70 percent of permanent employment and 80 percent of value-added in all modern sector industrial firms. It includes virtually all large firms and a great number of the small ones. It is hence designed to capture the entire size distribution of modern sector firms, but it naturally undersamples the numerous small enterprises. In terms of sectoral coverage, the sample includes 24 firms in the food industry,[1] 18 firms in the chemical and extraction industries, 13 firms in the mechanical (machinery and equipment) industry, 11 firms in the textile, clothing, and leather industry, 10 firms in the paper and wood industry, and 4 firms in other industrial groups

TABLE 4.3

Senegal: Total Employment of Permanent Workers by Industry

	1977	1982	1984
Food	13,188	13,722	13,961
Tobacco and Matches	546	466	518
Textiles, Clothing and Leather	6,513	5,000	5,157
Wood	936	268	304
Paper	275	404	206
Printing	402	371	496
Chemicals	1,695	2,796	3,138
Construction	829	753	544
Mechanical Industries	2,313	2,942	4,555
TOTAL	26,697	26,722	28,879

Source: Direction de la Statistique for 1977, 1982 and 1984.

---

1. The food industry is composed of the following sub-industries: sugar, oil, fish, flour, and other food.

(matches and tobacco, etc.). The sample covers the period from 1980 to 1985 but some data were not available for 1985. As a result, we present data for both 1984 and 1985 as two different endpoints that can be used in assessing the evolution of employment in the sample. We should also mention that our sample data tend, in fact, to be more accurate than the official data because firms occasionally do not distinguish clearly between permanent and temporary workers in their reports to the government.

As the data in Table 4.4 indicate, permanent employment in the sampled enterprises as a whole grew very slightly in the 1980s. Moreover, our sample data indicate that the discrepancies in the employment and output trends are not as striking as the official

TABLE 4.4

Senegal: Employment of Permanent Workers by Industry in the Enterprise Sample

	1980	1982	1984	1985
Food	9,980	10,181	9,886	10,072
Tobacco and Matches	557	466	512	476
Textiles, Clothing and Leather	4,540	4,641	4,568	4,655
Wood	236	205	230	211
Paper	297	401	371	374
Printing	93	191	115	113
Chemicals	1,279	1,690	1,673	1,455
Construction	597	634	474	449
Mechanical Industries	1,491	1,950	2,188	1,988
TOTAL	8,988	20,359	20,017	19,793

Source: Enterprise sample

data suggest. The percentage increases in employment within the chemical and mechanical industries, for instance, are much smaller than those found in the official industry statistics presented in Table 4.3. In recent (1982-1985) years, the chemical industry in fact employed fewer permanent workers, while employment in the mechanical industry rose slightly during this period. Our data also indicate that, contrary to the official data in Table 4.3, employment of permanent workers in the textile, clothing and leather industry actually increased during the 1980-1985 period. Finally, some of the discrepancies between the output and employment trends can be attributed to the differential use of temporary workers by industries over time. For instance, as Table 4.6 demonstrates, the food industry has significantly increased its reliance on temporary workers in the 1980s.

4.3.2 Employment of Temporary Workers

Because the population and hence the labor force have been growing rapidly, the overall stagnation of permanent industrial employment is clearly a cause for concern. One possible explanation of the phenomenon is that entrepreneurs have increasingly relied on temporary workers as a means of maintaining labor-input flexibility in the presence of rigid government enforcement of permanent employment. Tables 4.5 and 4.6 present data from the sampled firms on the number of temporary workers in various industries and their wage bills relative to the wage bills of permanent workers.

As Tables 4.4 and 4.5 show, the number of temporary workers employed in the industrial sector is approximately 40 percent of the number of permanent workers and seems to have remained fairly constant over the first half of the 1980s.[2] There is, however, some fluctuation, which supports the managers' statements that they are using temporary workers as a way to adjust their labor cost to cyclical market conditions.

The number of temporary workers is, of course, a good measure of how many individuals have at one time or another

---

2. The figures for the food industry exclude one flour mill, which appears to have reported unreasonable data (when compared to the other flour mills and other firms in the sample). Specifically, the numbers given were 13,200 temporary workers in 1980, 20 in 1982, 547 in 1983, 13,824 in 1984 and 700 in 1985.

TABLE 4.5

Senegal: Employment of Temporary Workers by Industry in the Enterprise Sample

	1980	1982	1984	1985
Food	2,107	5,490	8,001	4,615
Tobacco and Matches	62	241	0	77
Textiles, Clothing and Leather	519	704	629	752
Wood	1	9	5	17
Paper	51	12	61	73
Printing	28	53	51	40
Chemicals	5,140	312	212	226
Construction	83	139	101	86
Mechanical Industries	279	264	205	204
TOTAL	8,270	7,224	9,265	6,090

Source: Enterprise sample.

worked on a temporary basis. Although this number is a useful indicator of worker welfare, the extent of on-the-job training and the breadth of modern sector employment, it does not distinguish between short vs. long spells of temporary employment. Table 4.6 shows the relative use of permanent and temporary labor input by firms by presenting the labor cost of temporary workers relative to that of permanent workers in the sampled firms. The data indicate that the wage bill allocated to temporary workers in each industry is relatively small when compared to the wage bill of permanent workers--on the order of 5 percent to 12 percent for most industries. Hence, the actual number of days each temporary laborer is employed is quite small relative to the number of days worked by each permanent worker. However, there is some indication that the use of temporary workers has been increasing. The proportion of the temporary wage bill to the permanent wage bill doubled from 1980 to 1984. Moreover, since temporary workers receive few fringe benefits, their wage bill, relative to the wage bill of permanent workers, underestimates their relative use.

TABLE 4.6

Senegal: Ratio of Total Wage Bill of Temporary Workers
to Total Wage Bill of Permanent Workers
in the Enterprise Sample

	1980	1982	1984	1985
Food	0.055	0.129	0.197	0.163
Tobacco and Matches	0.055	.025	0.053	0.041
Textiles	0.062	0.118	0.068	0.110
Wood	0.112	0.080	0.048	0.049
Paper	0.039	0.006	0.050	0.053
Printing	0.121	0.069	0.193	0.103
Chemicals	0.046	0.025	0.031	0.032
Construction	0.078	0.115	0.149	0.116
Mechanical Industries	0.023	0.032	0.030	0.026
AVERAGE	0.051	0.087	0.116	0.103

Source: Enterprise sample

4.3.3 National Composition of Employment

Senegal's labor force is usually classified by nationality into Senegalese nationals (Senegalese), nationals of other African countries (other Africans), and citizens of non-African countries (non-Africans). Other Africans are for the most part nationals of other French speaking, West African countries. The majority of non-Africans are French citizens. The Syro-Lebanese also form an important economic group, although their impact is often underestimated in official statistics because many of Syro-Lebanese are Senegalese or French citizens.

Of major policy interest is whether the official policy of Senegalization has succeeded in increasing the employment share of the Senegalese nationals. Although growth in productive employment would be preferable overall, in its absence it is likely that an increase in employment of Senegalese nationals within a stagnant total labor force would be seen as a positive outcome from the national welfare standpoint.[3]

The first point to be noted is that temporary workers in the industrial sector appear to be predominantly Senegalese. This is clearly the case in enterprises that break down their temporary worker statistics by national groups. Other enterprises do not identify temporary workers by national group, but our interviews suggest that in these cases the temporary workers are also primarily Senegalese. Hence, as much as one-third of Senegalese employment in the industrial sector is characterized by temporary work and low pay (presumably reflecting the short average job duration and possibly lower wages).

The statistics in Table 4.7 indicate that the share of Senegalese in permanent industrial employment increased steadily from 87 percent in 1962 to 97.6 percent in 1985. During the same period the share of other Africans decreased from 4.7 to 0.7 percent, while that of non-Africans (primarily French) fell from 8.3 to 1.7 percent. These data reflect a significant substitution of Senegalese workers for non-Senegalese since independence. At the same time, the share of Senegalese clearly rise much more,

---

3. This of course is clearly the case if the skills of the displaced foreigners are identical or inferior to those of the incoming Senegalese workers. However, if the displaced foreigners are job-creating individuals (e.g., entrepreneurs), the process of Senegalization could result in a decrease of Senegalese employment from what it would have been in its absence. This issue is addressed in Chapter 6.

and future increases in the employment of Senegalese nationals will have to come from new employment. Unfortunately, aside from the 1980-1982 period, little permanent employment was generated for the Senegalese in the first half of the 1980s. As the data in Table 4.8 show, the number of permanent Senegalese workers in the sampled firms increased from 18,266 in 1980 to 19,653 in 1982 but has been declining slightly since then.

TABLE 4.7

Senegal: National Composition of Total Industrial Employment

	1963	1975	1980	1984	1985
Senegalese	87.0%	93.0%	96.0%	97.1%	97.6%
Other Africans	4.7%	2.3%	1.2%	0.9%	0.7%
Non-African	8.3%	4.7%	2.8%	2.0%	1.7%
TOTAL	100.0%	100.0%	100.0%	100.0%	100.0%

Source: Pfeffermann (1968) for 1963 and Svejnar (1984) for 1975. Enterprise sample for 1980, 1984, and 1985

TABLE 4.8

Senegal: Employment of Permanent Senegalese Workers by Industry in the Enterprise Sample

Industry	1980	1982	1984	1985
Food	9,546	9,852	9,618	9,837
Tobacco and Matches	541	453	500	465
Textiles, Clothing and Leather	4,424	4,523	4,460	4,594
Wood	222	192	219	200

Paper	265	371	354	358
Printing	85	183	108	105
Chemicals	1,210	1,617	1,606	1,404
Construction	570	606	453	430
Mechanical Industries	1,403	1,856	2,103	1,916
TOTAL	18,266	19,653	19,421	19,309

Source: Enterprise sample for 1980, 1982, 1984, and 1985.

Tables 4.9 and 4.10 depict the evolution of the skill (occupational) structure of the employed Senegalese, non-Africans, and other Africans, from 1975 to 1985. The Senegalese have been experiencing substantial upward mobility. Their

TABLE 4.9

Senegal: Number of Upper Level Managers by National Groups

Group	1975	1980	1984	1985
Senegalese	91	111	209	230
Other Africans	11	4	1	3
Non-Africans	280	237	200	190

Source: Svejnar (1984) for 1975. Enterprise sample for 1980, 1984, and 1985.

representation in the workers and apprentices category has decreased, and their numbers among the skilled workers, technicians, and middle management categories have increased significantly. As Table 4.9 demonstrates, the Senegalese have also become much more numerous among the managers, having replaced non-Africans as the dominant group in 1984. Non-Africans have virtually disappeared from the workers and apprentices group (Table 4.10), and they have also drastically reduced their presence among skilled workers, and technicians. About one-half (190 in 1985) of the remaining non-African falls into the top management category, and 38 percent (142 individuals) are top technicians and middle managers. Other Africans, whose presence is now minimal, are primarily found among the workers and technicians.

Tables 4.9 and 4.10 imply that there has been a significant upgrading of the permanently employed labor force, with the unskilled workers and apprentices becoming much less numerous over time. This is indeed the case, and it in part attests to an improvement in the quality of the permanent labor force, with temporary workers presumably being used more intensively in the unskilled worker group. However, part of the structural change is spurious because employers admit to yielding to worker and trade union pressure to reclassify workers even if these workers lack the corresponding skills. Part of the change, therefore, simply reflects reclassifications that correlate strongly with seniority but not necessarily with a corresponding increase in human capital. Finally, as we will discuss in Chapter 5, a major complaint about the functioning of Senegal's industrial labor market is the perceived decline in the provision of training (apprenticeships) by firms. If significant, this decline could explain the overall decrease in the number of individuals found in the workers and apprentices category.

4.4 Earnings

4.4.1 Components of Total Compensation

As we mentioned earlier, employee compensation in the modern sector usually consists of the base wage or salary (salaire de base), wage supplement (sursalaire), and various bonuses (primes). The employer also pays social security contributions, which are levied on the wage bill, and often also pensions. The base wage and some of the bonuses (e.g., those for seniority and transportation) are determined by the Tripartite Commission or through bargaining, whereas the wage supplement and some other bonuses (e. g., productivity rewards) are frequently determined unilaterally by the employer.

TABLE 4.10

Senegal: Composition of Permanent Employment by Skill Group

Group	1975	1980	1984	1985
A. Senegalese				
Top Management	0.9	0.5	1.0	1.1
Top Technicians and Middle Management	1.5	2.3	2.6	2.7
Technicians and Skilled Workers	16.5	16.8	21.5	22.0
Workers and Apprentices	81.1	80.2	74.9	72.2
TOTAL	100.0	100.0	100.0	100.0
B. Non-Africans				
Top Management	42.5	40.2	46.3	50.9
Top Technicians and Middle Management	23.4	34.0	36.0	38.1
Technicians and Skilled Workers	27.6	20.7	14.5	10.7
Workers and Apprentices	6.5	5.0	3.2	0.3
TOTAL	100.0	100.0	100.0	100.0
C. Other Africans				
Top Management	1.8	1.7	1.1	1.9
Top Technicians and Middle Management	2.3	3.1	3.9	4.5

(Continued)

Table 4.10 (Con't.)

Group	1975	1980	1984	1985
Technicians and Skilled Workers	23.8	13.5	23.8	16.0
Workers and Apprentices	71.1	81.7	71.3	77.6
TOTAL	100.0	100.0	100.0	100.0

Source: Svejnar (1984) for 1975. Enterprise sample for 1980, 1984, and 1985.

Although knowledge of the importance of base wages, supplements, and bonuses in total compensation is indispensable for understanding enterprise performance and formulation of public policy, no systematic information existed or could be easily obtained on this subject. Therefore, we administered a questionnaire to a stratified random sample of workers, technicians, and managers (612 in total) in a 17-firm subsample of our larger (80-firm) sample. Our analysis of earnings composition and determinants is based on information from these interviews.

As can be seen from columns 1 and 2 in Table 4.11, the ratio of base wage to total compensation decreases rapidly with increasing education and professional standing. As reflected by the employee survey data, the base wage forms 67 percent of a typical laborer's or unskilled worker's total compensation, but only 24 percent of the high level manager's total compensation. In order to provide a better understanding of this phenomenon, we calculated the proportions of total compensation that are comprised of bonuses and wage supplements, as shown in columns 3 and 4 of Table 4.11. The resulting proportions demonstrate that wage supplements, rather than bonuses, are primarily responsible for increasing the gap between base wage and total compensation as the employee's skill and education increase. Wage supplements represent 44.9 percent of the typical high level manager's total compensation but only 6.2 percent of the typical

TABLE 4.11
Senegal: Average Monthly Base and Total Compensation By
Skill Category in 17 Firms, 1986

Category	Compensation		Percent of Total Compensation Comprised of	
	Base	Total	Bonuses	Wage Supplements
High--level manager	126,593	520,09	18.85%	44.9%
Upper-level technicians and middle level managers	100,181	195,134	20.3%	15.3%
Middle-level technicians and lower level managers	78,289	160,198	25.8%	16.5%
Skilled employees	58,292	110,906	29.3%	11.0%
Skilled workers and unskilled employees	46,469	77,642	26.5%	7.9%
Laborers and unskilled workers	38,258	56,963	22.7%	6.2%

Source: Authors' survey of workers in 17 firms.

laborer's and unskilled worker's compensation.[4] In contrast, the variance in bonus percentages across skill categories is rather small--18 percent to 29 percent. Bonuses are, however, an increasingly important component of total compensation for the lower skill groups, rising up to one-third, leveling off, and then falling for the upper two skill groups.

---

4. Incidentally, the wage supplement is also creating a larger wedge between the average earnings of the top and the low level skill categories. Whereas the ratio of the base salary of high level managers to laborers and unskilled workers is 3.3, the same ratio using total compensation is 9.1.

Our interviews with enterprise managers also suggest that wage supplements have become more important as base wages began to fall relative to inflation. This development is important for two reasons: First, from an analytical standpoint we clearly must include supplements and fringe benefits in order to obtain a realistic picture of the evolution of worker earnings. Second, the preeminence of wage supplements in recent years has greatly decreased union influence on wages because many supplements are granted and adjusted at the employer's discretion.

4.4.2 Evolution of Wages by Skill Groups

Because base wages have been changed infrequently by the Tripartite Commission, the evolution of base wages has been very discontinuous. Percentage increases in base wages have tended to be related to the price increases of basic commodities, and the adjustments have generally favored low-income workers.

Data on the evolution of base wages suggest that, while in the past wages tended to catch up with inflation, since 1975 real wages have fallen significantly. For example, Bloch's (1985) calculation of the real minimum wage (real SMIG) suggests that it declined from an index value of 100 in 1975 to 90 in 1980 and 76 in 1984. Our calculations indicate that it decreased further, to 70 in 1985 and 66 in 1986.

Our data from the worker sample and data compiled by the Directorate of Labor tell the same story. The Directorate of Labor data on the evolution of the base wages of various categories of workers between November 1974 and April 1983 indicate that the cumulative percentage increase ranged between 70 and 83 percent for the unskilled workers and between 35 and 45 percent for the professionals of the highest category. During the same period, the African and European cost of living indices registered increases of 107 and 114 percent, respectively, thus substantially exceeding the rise in the base wages of all categories.

The survey data on workers in 17 enterprises, presented in Table 4.12, indicate that average annual growth of the total compensation (measured as base wage plus bonuses and supplements) in nominal terms was 6.55 percent a year over the

1984-1986 period.[5] The average annual growth of the base wage over these years was 4.18 percent. Given that the rate of inflation (as measured by the cost of living index) was approximately 13 percent a year from 1984 to 1986, it is clear that the industrial wage (in terms of the base or total) has not been keeping up with the cost of living.

Table 4.12 also examines the growth of total compensation by its principal components within the major skill categories found

TABLE 4.12

Senegal: Average Annual Percentage Growth of Nominal Monthly Compensation by Skill Category, 1984-1986

Category	Base Wage (%)	Bonuses (%)	Supplements (%)	Total Compensation (%)
High level manager	6.87	3.08	51.80	
Upper-level technicians and middle level managers	5.35	7.11	16.43	17.76
Middle-level technicians and lower level managers	2.89	8.51	15.59	
Skilled employees	1.96	8.53	11.16	5.48
Skilled workers and unskilled employees	2.63	10.16	8.19	
Laborers and unskilled workers	7.28	14.11	6.33	3.50
Average Annual Growth	4.18	9.90	12.88	6.55

Source: Survey of workers in 17 firms.

---

5. This time period is, of course, a short time period for drawing conclusions about wage trends. However, because the data come from worker surveys, they represent a useful complement to our other data.

in the 17 firm sample. Our data from the 17 firm sample give a somewhat different picture than those of the Directorate of Labor in that (during the 1984-1986 period) the average growth of the base wage of laborers and unskilled workers is not much (if at all) higher than that of high level managers. This difference in findings suggests that occupational differences in wage growth may have moderated in recent (1984-1986) years in comparison to the earlier (1974-1983) period.

Nevertheless, from Table 4.12 it is clear that, because of the relatively rapid growth of the supplements of the two higher skill groups (especially the high-level managers), the average growth of total compensation has in recent years been substantially higher for the upper level technicians and middle to high level managers than for the two lowest skill groups--17.76 percent a year compared to 3.5 percent a year. Hence, our data from the subsample of 17 firms suggest that in the 1984-1986 period, the total compensation of the highest category of labor was keeping abreast of the cost of living while that of the low level worker was falling behind.

4.4.3 *Evolution of Total Compensation by National and Skill Groups*

Table 4.13 presents data from the large (80-enterprise) sample on the average compensation of the three National groups within four skill categories in 1980, 1984, and 1985. The data reflect the total compensation of employees (including employer social security contributions) as reported by employers in the 80 sampled firms. In contrast to the 17-firm sample, these data come from employer records and they cover a longer time period as well as a larger group of employees. Like the data from the 17-firm sample, the 80 firm data in Table 4.13 indicate that total employee compensation has not risen as rapidly as the cost of living in the 1980s.

Table 4.13 also provides information about differentials in total compensation among the three national groups. The table demonstrates that the growth in the compensation of different categories of workers was not uniform across the national groups. The compensation of non-African employees has outstripped that of Senegalese and other African employees in almost every category and has thus slightly increased the substantial differentials between the compensation of non-Africans and

Africans. For example, the average compensation of non-African top managers rose by 65 percent, whereas that of Senegalese top managers rose by 59 percent, over the 1980-1985 period. This differential growth rate increased the already sizable ratio of the average compensation of non-African to Senegalese top managers from 2.5 to 2.6. While comparable data for earlier periods are not readily available, it should be noted that the above evolution in relative managerial compensation is probably quite favorable for the Senegalese by historical standards.

TABLE 4.13

Senegal: Average Annual Compensation Per Employee by National and Skill Groups

A. 1980	SENEGALESE	OTHER AFR.	NON-AFR.
Top Technicians	5,013	10,210	12,595
Top Technicians and Middle Management	2,767	2,268	9,167
Technicians and Skilled Workers	1,235	1,232	5,949
Workers and Apprentices	581	656	774
B. 1984	SENEGALESE	OTHER AFR.	NON-AFR.
Top Management	7,751	6,774	19,199
Top Technicians and Middle Management	4,125	4,707	12,943
Technicians and Skilled Workers	1,680	1,665	9,062
Workers and Apprentices	855	833	655

(Continued)

Table 4.13 (Con't.)

C. 1985

	SENEGALESE	OTHER AFR.	NON-AFR.
Top Management	7,972	6,130	20,812
Top Technicians and Middle Management	4,474	4,431	15,437
Technicians and Skilled Workers	1,677	1,993	11,624
Workers and Apprentices	950	855	410

Values are in 1,000 of Francs CFA.
Source: 80 Enterprise sample

The more striking case is the growth in average compensation of non-African technicians and skilled workers (95 percent) as compared to that of similarly skilled Senegalese (36 percent) and other Africans (62 percent). In this category of workers, this diversity in compensation increases led to a jump in the ratio of the non-African earnings to Senegalese earnings from 4.8 to 6.9 and in the ratio of non-African to other African earnings from 4.8 to 5.8 between 1980 and 1985. Hence, although Senegalization in the labor market has taken place in terms of increased employment of the Senegalese in more skilled positions, the gap between the per employee compensation of non-Africans and Senegalese has been widening. It should be noted, however, that because of the substantial exodus of the non-Africans, the mean compensation per employee in the two lowest skill groups may be quite sensitive to the characteristics of the few remaining non-African workers. Moreover, an evaluation of the merits of the emerging earnings differentials would require information on the relative productivity of individuals in the three national groups.

5

Labor Market Regulations, Industrial Relations, and Enterprise Performance

In this chapter we analyze qualitatively the effects of labor regulations and the industrial relations system on enterprise performance. Our approach is institutional, and it relies primarily on the information gathered during our interviews with the government officials, employers and the representatives of their associations, and workers and their trade union representatives.

5.1 The Position of Government Officials

The government officials that are most directly involved in the labor market are those in the Ministry of Public Affairs and Labor. Within this ministry, they are located primarily in the Directorate of Employment and the Directorate of Labor. The staff in these directorates is composed of professionals with legal and various social science backgrounds. It includes researchers, administrators, and labor inspectors. The labor inspectors are in charge of examining employment and compensation practices in individual firms and ensuring that they conform to the Labor Code and other regulations.

Because philosophically the government is in a loose alliance with the cooperating trade union, CNTS,[1] most government officials are in the position of trying to promote an environment that allows for the protection of workers' interests without endangering the existence of firms. Until very recently, the government relied on labor market regulations as an essential means for furthering this goal. The interventionist approach

---

1. There are also more radical trade unions that consider collective bargaining in the tripartite framework to be inconsistent with class struggle and the true mission of trade unionism. These unions denounce CNTS and refuse to cooperate with the government. See also Martens (1982).

followed naturally from the French system and has been perpetuated through recruitment of Ministry staff primarily from the legal profession. With the onset of the New Industrial Policy in the mid-1980s, the government has gradually searched for ways of alleviating the regulatory burden with the double aim of enhancing enterprise flexibility and preserving workers' welfare.

In practice, government policy has been reflected in the numerous regulations and recent deregulations discussed in Chapter 3. At present, government officials are still apprehensive about undertaking major reforms of the industrial relations system, although they have concluded a major study in this area under a World Bank contract.[2] Discussions with officials at all levels revealed a continuing mistrust of enterprise policies and a fear that reforms might disrupt the fragile balance in the adversarial union-management relationship.

As regards specific regulations on hiring, employment contracts, layoffs, and wage setting, most government officials now feel that hiring need not be tightly regulated and that direct matching between workers and firms should be encouraged. However, they maintain that enterprises now have sufficient flexibility with respect to employment contracts and layoffs and that loosening of the current rules would have a negative impact on worker welfare. In particular, officials claim that the use of temporary workers provides sufficient flexibility for firms and that government processing of employer requests for layoffs is rapid and frequently leads to approval. In addition, most officials see the existing minimum wage regulation as basically ineffectual because the cost of living has in recent years risen faster than the SMIG and because many salary supplements continue to be determined unilaterally by the employer or bilaterally between the employer and the union. In sum, the government officials dealing with labor issues believe that the burden of improving the performance of Senegal's enterprises falls on neither the government nor the workers, but rather on the managers.

---

2. See Delegation a l'Insertion a la Reinsertion et a l'Emploi (1988).

5.2 Managers' Views

Our interviews with employer representatives and managers in the 17 firms in our sample indicate that the vast majority of them are dissatisfied with the Labor Code and the system of labor relations in Senegal. Most managers feel that the regulations prevent them from making efficient decisions and, more importantly, that they reduce the motivation and productivity of the workers. They regard the low levels of education, discipline, motivation and productivity of most workers as detrimental to enterprise performance and to the economic development of Senegal in general. Although these sentiments were perhaps stronger among the expatriate and Syro-Lebanese managers, the Senegalese managers shared these views as well.

5.2.1 Restrictive Labor Regulations

As part of our interview process, we asked the managers to identify the labor regulations that most restricted their economic activities. As shown in Table 5.1, nearly two-thirds of those managers surveyed identified the regulations on layoffs as being the most binding. Also mentioned, but by 15 percent or less of the sample, were: 1) the regulations according daily workers permanent status if employed for more than four consecutive days; 2) limits imposed on the length of contract that can be offered to new expatriate employees; and 3) regulation of overtime work. Finally, the functioning of the Labor Service of the Ministry of Public Affairs and Labor was also mentioned as an important limitation on enterprise activity.

In discussing their problems, most managers expressed the feeling that in many cases the Labor Service was more a nuisance than a constraint. Almost all indicated that they often hire workers they want and only subsequently "regularize" the workers' status with the Labor Service; however, the process was not automatic and sometimes employers were required to interview four or five people sent by the service before they could hire the workers they wanted.

In order to assess the extent to which regulations affect flexibility in labor utilization, we asked the managers to identify methods of adjusting their labor cost and the number of hours worked during the business cycle. Sighing and throwing up of hands was the predominant first response. Some managers indicated that they have no way of adjusting because the workweek is fixed by the Labor Code at forty hours. Most said that they use daily workers as a cushion or that they vary the number of hours of overtime. Only one manager said he used subcontracting as a means of adjustment.

TABLE 5.1
Senegal: Most Constraining Labor Regulation According to Managers

Constraint	Percent of Total Sample
Regulations on Layoff	61%
Regulation of Daily Workers	15%
Regulation on Hiring Expatriates	15%
Labor Service	8%
Regulation of Overtime Work	8%

Source: Authors' survey of managers, 1986.

Are wages being set too high (or too low) relative to labor productivity? Most of the managers in our sample did not feel that wages reflect worker productivity. However, there is some consensus that the wage of the expatriates is either equal to or lower than their level of productivity and that the wages of unskilled workers are higher than or equal to their productivity.

When asked whether they would hire more labor if workers' salaries were reduced by, for example, 20 percent, 91 percent of the managers responded that they would not. Some managers also added that the wages of unskilled workers should not be lowered, as they are already at the "socially acceptable minimum" in terms of basic needs. Many stressed that the problem is not high wages but rather poor effort, lack of worker discipline, and low productivity.

5.2.2 Recruitment and Training

We asked several questions about recruitment and learned that, in general, it is not difficult to hire the unskilled and semi-skilled

skilled workers at current salaries; it is, however, difficult to hiring recruit the higher-skilled Senegalese and non-African employees. Managers cite as the principal obstacles the shortage of well-trained Senegalese and the constraints of existing regulations that make it less attractive for expatriates to accept employment. Salaries are not considered to be a problem. Interestingly, similar results are reported by Pfeffermann (1968) from his 1963 survey of the managers in 22 firms: When asked what was the major problem with the African labor force, 45 percent of the managers interviewed by Pfeffermann answered "qualifications."

An important point to note is that whereas most employers complain about the lack of skills in the labor force, relatively few provide training. We found that only one-third of the workers in the 17 firms we surveyed had received some training in the firm. Comparing this figure with Pfeffermann's finding that 42 percent of the firms in 1963 had manpower training schemes (Pfeffermann, 1968, p. 285), it appears that the amount of training provided by most firms has been declining. It is of course impossible to draw a firm conclusion in this respect because of the potential differences in the two samples.

In assessing the perceived lack of skills in the labor force, one of course has to take into account the extent and quality of the educational system. Available data indicate that the education budget has increased from less than 1 percent of GDP in 1960 to 5 percent of GDP in 1984. While this increase is impressive, the net enrollment rate in primary education is still only 40 percent and is expected to decline in view of the containment of public expenditures and the 2.9 percent annual population growth rate. This, and the fact that only 40 percent of students entering primary schools graduate, suggests that the government's objective to achieve universal primary education by the end of the twentieth century may be too ambitious. In addition to suffering from low enrollments, Senegal's system of primary education is reported to be exceedingly academically oriented, at the expense of practical training which might facilitate access to productive employment. Our interviews with workers revealed many cases of illiteracy, thus also providing evidence that the educational background of many workers is very limited.

As mentioned previously, it appears that many workers were originally recruited not through the Labor Service but through other channels. The managers simply "regularized" the workers with the Labor Service after they had hired them. Procedures for recruiting vary by skill level: The most common procedure for hiring unskilled workers (ouvriers-maneuvre) appears to be to recruit them at the gate as daily workers and offer them permanent contracts if they perform well. As seen in Table 5.2, over two-thirds of the managers we surveyed indicated they follow this practice. In observing the number of men waiting at the gates, we could see that the supply of daily workers is unlimited in the relevant range. Workers are sometimes hired in this manner because they are recommended by someone who works in the firm.

TABLE 5.2
Senegal: Methods of Recruiting
Employees at Various Skill Levels
(Manager Questionnaire)

Method of Recruiting	Unskilled Workers	Skilled Workers	Technical Personnel	Managers
Hire First as Daily Workers	69 %	-	-	-
Ask Workers for Recommendation	15 %	15 %	8 %	-
Promotion Within the Firm	-	23 %	15 %	-
Advertise in Newspaper	15 %	24 %	8 %	-
Hire Directly from School	-	8 %	8 %	-
Hire from Other Firms in:				
Senegal	-	-	8 %	15 %
Europe	-	-	-	15 %

Note: These figures represent percentages of the total sample of managers (17) that gave these answers. Answers were not provided by some managers for some skill groups.
Source: Authors' survey of managers, 1986.

Skilled workers and white collar employees are generally hired through newspaper advertising or directly from schools or other firms. Internal promotion also plays an important role. Although most firms tend to hire their technical personnel from other firms, some technicians are acquired through internal promotion. Finally, managers are primarily sought from other firms, both in Senegal and in Europe.

A recent study of the employment practices in Senegal's state owned firms was performed by Lachaud (1987). In his detailed investigation, Lachaud finds that the access to public sector jobs varies across occupations: education appears to be the important prerequisite for higher level positions, while personal contacts appear to be the principal avenue for obtaining less skilled jobs.

5.3 Workers' Views of Unions

The vast majority of Senegal's industrial workers are unionized. Within our sample, 83 percent of all workers belong to a union; in 1963, within Pfeffermann's sample, the figure was 72.8 percent (See Pfeffermann [1968] Table 39, p. 301). However, these numbers do not accurately reflect the strength and effectiveness of the unions as seen from the workers' perspective. When asked "What are the main advantages of union membership?" the responses of workers were unenthusiastic and sometimes very critical. Judging from their expressions, most workers feel that the unions are not doing very much. Even so, workers were unlikely to criticize the unions when given the opportunity; in our questionnaire most of them gave the answers they felt "should be given." In trying to discover the reason for this conformity, it appears to us that few workers criticized the unions because the unions are closely aligned with the government and our questionnaire, although guaranteeing respondent anonymity, was officially approved by the government census office.

The worker responses which we obtained are consistent with those from the 1963 survey by Pfeffermann (1968, pp. 301-304) in which similar questions were asked. To the question "What are unions there for?" 71.3 percent of the workers in 1963 identified "protection against bosses and defense of interests," which is strikingly similar to the 69.7 percent of our sample that identified some form of union protection (see Table 5.3). When asked about their general opinion of unions, 48.5 percent of the workers interviewed in 1963 had no opinion and 31.7 percent criticized unions. (About 20 percent did not answer this question because they were not union members.) When asked "Who initiates strikes?" 52.3 percent of the workers interviewed in 1963 answered "union officials" and 9 percent answered "the workers." When asked to identify the main criticism of their own union, 59.5 percent of these workers answered "gap between leaders and workers," 21 percent said "corruption," and 6.3 percent noted "excessive political involvement."

TABLE 5.3
Senegal: Advantages of Union Membership

	Percent of Total Who Volunteered This Answer:
Protection of:	
Wages	41.7 %
Working Conditions	21.0 %
Other	7.0 %
No advantages	17.7 %
No opinion	10.0 %
No answer	3.6 %

Source: Authors' survey of workers, 1986.

Combining the information from the 1963 and 1986 surveys of the workers' opinions, we obtain the impression that trade unions in Senegal are, and have been, generally aligned with powerful political groups and somewhat divorced from the workers' interests. This view of the "cooperating" unions is also held by the the noncooperating (opposition) trade union organizations. The question that naturally arises in this context is why the majority of workers join the unions. The answer seems to lie in low union dues and a sense of protection against potentially extreme actions by employers.

5.4 Trade Union Perspective

On the several occasions that we met with the leaders of the CNTS, we were struck by their low-budget operation, their adversarial rather than cooperative approach to industrial relations, and the fact that although some of their positions were based on very accurate information, others were founded on a total lack of information and accuracy.

The CNTS operates in an alliance with the government and in severe mistrust of the employers. It attributes Senegal's economic problems to the lack of investment, to poor labor policies, and to limited worker training within most firms. CNTS sees its activities as essential for the protection of workers' welfare and as not harmful to the efficient operation of enterprises. Interestingly, trade union mistrust of employers does not appear to be abating with the Senegalization of management (see Table 4.9), but rather appears to be based on class lines.

5.5 The Noncooperative Outcome

The most important feature of the highly adversarial system of industrial relations is the economic and social outcome, which is suboptimal from the standpoint of all the concerned parties. The trade unions' severe mistrust of management results in union pressure on the workers to resist exploitation by the employer. This translates into low worker motivation and a decrease in overall labor productivity. Moreover, because job security and the base wage are virtually guaranteed by labor regulations and because the total wage is not directly linked to performance, lessened effort and low productivity are to be expected. Job security regulations prevent employers from paying efficiency wages and induce them to substitute daily (temporary) workers for the permanent ones as often as possible. The government has attempted to limit these practices by restricting the use of temporary workers and by permitting layoffs only in extreme cases. This, in turn, has discouraged employers from hiring new permanent workers and has possibly increased turnover of temporary (unskilled) workers (see, e.g., the 1985 figures in Table 4.5). The outcome appears to be a low productivity, low wage, noncooperative solution. As we show in Chapter 6, this institutionally based hypothesis is consistent with the econometric findings of falling total factor productivity.

6

The Productivity Effects of Labor Force Structure, Enterprise Ownership, and Foreign Trade Variables

In this chapter we analyze quantitatively the determinants of productivity in the 80 sampled firms. The data set is in the form of an annual panel covering 80 firms during the 1980-1985 period. Our focus is on the evolution of total factor productivity over time and the productivity effects of the following variables: (1) the composition of the enterprises' labor force (in terms of skill, nationality and permanent versus temporary employment); (2) enterprise ownership, as determined by the percentage of enterprise shares owned by Senegalese citizens, the Senegalese state, or foreign individuals and companies (in 1984 or 1985); (3) the effective rate of protection and enterprise export orientation.

As we pointed out in Chapter 5, employers often complain that the low productivity of Senegal's workers is one of the main reasons for the poor performance of the country's industrial sector. Employers also argue that the enormous wage differentials between Senegalese and non-African employees are (more than) justified by their different productivities. In contrast, the trade unions argue that such differentials are frequently unwarranted and constitute a siphoning-off of profits that ought to be reinvested or used for the remuneration of Senegalese workers.

These arguments have been aired for decades, but the issue has remained unsettled. To provide a more solid background for examining this question, we have collected detailed firm-level data on production, labor cost, and employment variables, and on a variety of enterprise and worker characteristics. With these data, we estimate augmented production functions and assess the extent of overutilization of different types of labor as well as the effect of the aforementioned structural and policy variables on output.

Our general approach is to estimate industry-specific and overall (sample-wide) production functions of the form

$$\ln Q = g(Z)f(X),$$

where Q is value added in constant domestic or world prices, X is a vector of inputs, and Z is a vector of structural and policy variables that may affect productive efficiency and lend themselves to or become potential tools in a policy reform. The function f(X) is hence the input function, and g(Z) is the productive efficiency function.

The X variables that we use include the gross capital stock, the number of permanent workers in the various national and skill groups,[1] and the deflated wage bill of temporary workers. The latter variable is used as a measure of the temporary labor input. Apart from the time trend, which captures the evolution of productive efficiency over time, the Z variables include enterprise ownership measured by the percentage of enterprise capital owned by (a) Senegalese state--referred to in the tables as "% Senegalese State," (b) French citizens or companies--referred to as "% French," (c) Syro-Lebanese nationals or companies-- referred to as "% Syro-Lebanese," and (d) individuals, companies or governments of other countries--referred to as "% Other Foreign." The Z vector also includes the percentage of output that is exported (capturing efficiency differentials between exporting and nonexporting firms), and the effective rate of protection (reflecting productivity differences between more and less protected firms).[2]

The latter three variables (ownership, export orientation, and effective protection) are of particular relevance to public policy. Although econometric evidence is scarce,[3] institutional knowledge has led numerous governments and international organizations,

---

1. In the few instances when no workers of a given nationality were found in a given group (cell), we set the value of the natural logarithm of this labor input variable equal to 0.001. The results are not sensitive to variations in this number.

2. The effective rate of protection is defined as the percentage differential between a firm's value-added calculated with domestic and world prices.

3. See for example Aharoni (1986) and Svejnar and Hariga (1987) for a discussion of the existing studies.

such as the World Bank, to pursue policies that favor private ownership, export orientation, and decreased trade protection. At the enterprise level, these policies are motivated by the belief that private, export-oriented and less protected firms tend to be more efficient than others. Within Senegal, there is also a widespread belief that foreign-owned firms tend to be more efficient than domestic-owned ones. Our data and analytical design enabled us to test these hypotheses.

In estimating the production functions, we have tried several specifications: an augmented Cobb-Douglas form, an augmented Cobb-Douglas form with linearly additive components within each skill group, and more flexible forms such as the translog. The Cobb-Douglas function has numerous desirable properties, including parsimony and ease of interpretation, but its principal drawback is its assumption that all inputs can be substituted with a unitary elasticity. This may not be totally realistic but, because at the most disaggregated level we are dealing with thirteen different labor inputs,[4] the use of more flexible functional forms (e.g., translog) proved too cumbersome.

In our view, the main problem with the Cobb-Douglas function is that it may underestimate the ease with which firms can substitute different nationals with the same skills (e.g., skilled Senegalese and skilled other-African workers) in production. It could be argued that these workers have different productivities but are very easily substituted for one another. To allow for this possibility, we have estimated the Cobb-Douglas function with the labor input of different national groups entered in a linearly additive form within each skill group. This amounts to assuming that the elasticity of substitution of all workers within the same skill group is infinite and reflects better the hypothesis that substitution of different nationals within skill groups is easy. Moreover, it permits a direct comparison of worker productivity across national groups. Unfortunately, the estimated parameters of relative productivities converged to implausibly large values and, when constrained to remain within reasonably wide bounds (e.g., a 20/1 maximum relative productivity differential between non-Africans and Senegalese within a given skill group), they invariably converged to the boundary. As a result, we have

---

4. These include four skill categories for each of the three national groups and the temporary workers.

invariably converged to the boundary. As a result, we have adopted the standard, augmented Cobb-Douglas function as our principal specification.

Tables 6.1-6.13 contain our estimates of various specifications of the Cobb-Douglas value-added functions. Tables 6.1-6.3 report sample-wide estimates with industry-specific fixed effects, and Tables 6.4-6.13 contain parameter estimates by industry. Each table gives results based on deflated value-added in domestic and world prices. The world price data on value-added are based on the calculations of the effective rate of protection that were computed for each firm in our sample by a World Bank financed project carried out at the Senegal's Ministry of Finance. The availability of value-added data in both domestic and world prices enables us to assess the effects of the numerous variables in the context of both domestic distortions and the more objective scarcities of the world market.

6.1 Overall Production Function Estimates

Tables 6.1 and 6.2 report the instrumental variable (IV) and ordinary least squares (OLS) estimates, respectively, of three major specifications of the Cobb-Douglas production function. In each table, columns 1-3 contain estimates based on value-added in domestic prices; estimates in columns 4-6 are based on value-added in world prices. The instruments used are canonical correlation variables derived from firm-specific dummy variables interacted with time and with the value-added deflator (see e.g., Bowden and Turkington, 1984). The variables which we instrumented are the labor input variables, capital stock, and the percentage of output that is exported. The IV and OLS estimates yield similar results, with the OLS estimates being occasionally more significant.

As the estimated coefficients in Tables 6.1 and 6.2 indicate, total factor productivity appears to be unrelated to enterprise ownership, except for the Syro-Lebanese firms, which display a significantly (30-50 percent) inferior level of productivity. The results challenge three traditional notions, namely that (1) private firms are more productively efficient than public ones; (2) foreign-owned firms are more productively efficient than domestically-owned ones; and (3) Syro-Lebanese enterprises are more productively efficient than domestically-owned firms and, possibly, other foreign-owned firms. The results hold for various specifications based on value-added in both domestic and world prices, and the question naturally arises as to whether systematic misrepresentation of accounts could give rise to these findings.

The negative productivity effect of the Syro-Lebanese ownership is particularly intriguing, given the reputation for success on the part of the Syro-Lebanese entrepreneurs. In this context it is worth noting that our earnings function estimates in Chapter 7 indicate that Syro-Lebanese firms tend to pay seemingly comparable workers significantly (30-70 percent) less than do other firms. The productivity differential is hence consistent with a labor cost differential in these firms.

An important aspect that is often overlooked in the discussions on relative performance of private and public (government-owned) firms is the part played by the market structure (extent of competition). In particular, the relatively inferior performance of public firms is frequently observed in circumstances when these firms are accorded a much greater degree of market power or access to resources (e.g., subsidies) than their private counterparts. The few studies that have examined the two types of firms in competitive environments often do not find differences in performance.[5] In Senegal, the market structure is very concentrated and many private firms have succeeded in obtaining concessions (in the form of so called *conventions*) from the government. Some of these private firms (e.g., the sugar complex) are notorious for their inefficiency but they survive because of the special arrangements. This institutional and feature offers a possible explanation for our findings about the impact of enterprise ownership on productivity.

The estimated coefficients on the trade variables indicate that enterprise export orientation (measured by exports as a proportion of total sales) is unrelated to total factor productivity, but that the impact of effective protection is sizable and negative when value-added is expressed in world prices. The latter finding strongly supports the hypothesis that protection has a deleterious effect on productive efficiency, and it shows that the effect may not be detectable when distorted domestic prices are used to calculate value-added. The insignificance of the export orientation variable suggests that exporting per se need not induce the firm

---

5. See, e.g., Caves and Christensen (1980).

TABLE 6.1

Senegal (All Industries): IV Estimates of Augmented Cobb-Douglas Production Functions — Dependent Variable = Log of Constant Price Value Added

(Values in Parentheses are Standard Errors)

	Based on Value-Added in Domestic Prices			Based on Value-Added in World Prices		
	1	2	3	4	5	6
Intercept (% Senegalese Private)	1.836* (0.631)	1.183* (0.568)	-0.066 (0.512)	0.532 (0.681)	0.352 (0.609)	-0.824 (0.550)
Ownership:						
% Senegalese State	0.317 (0.278)	0.143 (0.254)	-0.063 (0.263)	0.169 (0.296)	0.280 (0.266)	-0.095 (0.272)
% French	0.094 (0.112)	0.044 (0.097)	0.119 (0.100)	0.002 (0.118)	-0.076 (0.103)	-0.003 (0.105)
% Syro-Lebanese	-0.455* (0.127)	-0.471* (0.112)	-0.452* (0.127)	-0.382* (0.133)	-0.516* (-0.117)	-0.445* (0.129)
% Other Foreign	0.028 (0.133)	-0.032 (0.119)	0.056 (0.120)	-0.086 (0.140)	-0.167 (0.123)	-0.076 (0.124)

Trade Variables:						
Effective Rate of Protection	-0.012 (0.007)	-0.010 (0.007)	-0.009 (0.007)	-0.352* (0.022)	-0.351* (0.021)	-0.320* (0.021)
% Exports	-0.159 (0.194)	-0.003 (0.178)	-0.045 (0.182)	-0.137 (0.208)	-0.135 (0.187)	-0.198 (0.190)
Inputs:						
ln (Senegalese Unskilled)	0.361* (0.056)	0.362* (0.055)		0.394* (0.060)	0.392* (0.058)	
ln (Other Africans Unskilled)	6.9E-6 (7.6E-5)			-8.7E-5 (8.3E-5)		
ln (Non-Africans Unskilled)	4.8E-4* (1.8E-4)	4.2E-4* (1.7E-4)		7.7E-4* (2.0E-4)	6.8E-4* (1.9E-4)	
ln (Senegalese Skilled)	2.5E-4+ (1.5E-4)	3.2E-4* (1.4E-4)		4.1E-4* (1.5E-4)	4.0E-4* (1.5E-4)	
ln (Other Africans Skilled)	2.7E-4* (9.5E-5)	2.7E-4* (8.4E-5)		2.9E-4* (9.9E-5)	2.2E-4* (8.6E-5)	
ln (Non-Africans Skilled)	1.2E-4 (9.6E-5)	3.5E-5 (1.0E-4)				

(Continued)

TABLE 6.1 (Con't.)

	Based on Value-Added in Domestic Prices			Based on Value-Added in World Prices		
	1	2	3	4	5	6
ln (Senegalese Technicians and Middle Managers)	1.9E-4* (8.2E-5)	2.0E-4* (7.5E-5)		1.7E4* (8.7E-5)	1.8E-4* (7.8E-5)	
ln (Other Africans Technicians and Middle Managers)	4.3E-4* (1.3E-4)	5.1E-4* (1.2E-4)		3.5E-4* (1.4E-4)	3.7E-4* (1.3E-4)	
ln (Non-Africans Technicians and Middle Managers)	1.5E-4+ (8.6E-5)			-5.1E-5 (9.1E-5)		
ln (Senegalese Upper Managers)	1.2E-4 (8.2E-5)			9.7E-5 (8.6E-5)		
ln (Other Africans Upper Managers)	1.9E-4 (1.9E-4)			9.3E-5 (1.9E-4)		
ln (Non-Africans Upper Managers)	-6.6E-5 (9.8E-5)			-9.7E-5 (1.1E-4)		
ln (Total Senegalese)			0.538* (0.066)			0.637* (0.070)

	(1)	(2)	(3)	(4)	(5)	(6)
ln (Total Other Africans)			2.0E-4* (7.3E-5)			3.8E-5 (7.7E-5)
ln (Total Non-Africans)			-1.9E-5 (1.6E-4)			-1.0E-4 (1.6E-4)
ln (Wage Bill of Temporary Workers)	1.9E-4* (8.6E-5)	1.8E-4* (7.9E-5)	1.7E-4* (8.3E-5)	2.2E-4* (9.2E-5)	1.6E-4+ (8.5E-5)	1.7E-4+ (8.7E-5)
ln K	0.479* (0.048)	0.510* (0.045)	0.450* (0.052)	0.572* (0.052)	0.571* (0.048)	0.484* (0.055)
Time	-0.048* (0.016)	-0.060* (0.016)	-0.056* (0.018)	-0.074* (0.017)	-0.074* (0.017)	-0.069* (0.017)
Industry Intercepts	yes	yes	yes	yes	yes	yes
R^2	0.883	0.882	0.879	0.881	0.881	0.880
N	387	387	387	376	376	376

* = Significant at the 5 percent statistical test level.
+ = Significant at the 10 percent statistical test level.
Source: 80-firm sample

TABLE 6.2

Senegal (All Industries): OLS Estimates of Augmented Cobb-Douglas Production Functions—Dependent Variable = Log of Constant Price Value Added

(Values in Parentheses are Standard Errors)

	Based on Value-Added in Domestic Prices			Based on Value-Added in World Prices		
	1	2	3	4	5	6
Intercept (% Senegalese Private)	1.347* (0.559)	0.865+ (0.519)	-0.042 (0.481)	0.171 (0.606)	0.062 (0.557)	-0.831+ (0.517)
Ownership:						
% Senegalese State	0.264 (0.256)	0.145 (0.241)	-0.016 (0.244)	0.158 (0.269)	0.242 (0.252)	-0.049 (0.252)
% French	0.115 (0.107)	0.089 (0.095)	0.132 (0.097)	0.016 (0.112)	-0.038 (0.099)	0.006 (0.101)
% Syro-Lebanese	-0.498* (0.121)	-0.489* (0.111)	-0.469* (0.119)	-0.473* (0.127)	-0.541* (0.115)	-0.486* (0.122)
% Other Foreign	0.047 (0.127)	0.005 (0.116)	0.062 (0.118)	-0.079 (0.134)	-0.138 (0.121)	-0.079 (0.021)

	(1)	(2)	(3)	(4)	(5)	(6)
Trade Variables:						
Effective Rate of Protection	-0.009 (0.007)	-0.008 (0.007)	-0.008 (0.007)	-0.342* (0.021)	-0.344* (0.021)	-0.322* (0.021)
% Exports	-0.112 (0.160)	-0.024 (0.154)	-0.027 (0.156)	-0.095 (0.169)	-0.098 (0.161)	-0.119 (0.163)
Inputs:						
ln (Senegalese Unskilled)	0.371* (0.051)	0.377* (0.050)		0.403* (0.054)	0.399* (0.053)	
ln (Other Africans Unskilled)	4.5E-5 (6.3E-5)			-5.8E-6 (6.7E-5)		
ln (Non-Africans Unskilled)	2.5E-4+ (1.4E-4)	2.33E-4+ (1.4E-4)		4.3E-4* (1.6E-4)	4.1E-4* (1.5E-4)	
ln (Senegalese Skilled)	1.8E-4 (1.2E-4)	2.3E-4+ (1.2E-4)		3.0E-4* (1.3E-4)	2.9E-4 (1.3E-4)	
ln (Other Africans Skilled)	2.0E-4* (7.6E-5)	2.2E-4* (7.2E-5)		2.0E-4* (7.9E-5)	1.9E-4* (7.4E-5)	
ln (Non-Africans Skilled)	8.6E-5 (7.7E-5)			-5.7E-6 (8.1E-5)		
ln (Senegalese Technicians and Middle Managers)	1.9E-4* (7.3E-5)	1.8E-4* (6.9E-5)		1.6E-4* (7.7E-5)	0.164E-3* (0.717E-4)	

(Continued)

Table 6.2 (Con't.)

	Based on Value-Added in Domestic Prices			Based on Value-Added in World Prices		
	1	2	3	4	5	6
ln (Other Africans Technicians and Middle Managers)	3.8E-4* (1.1E-4)	4.2E-4* (1.0E-4)		3.1E-4* (1.1E-4)	0.308E-3* (0.106E-3)	
ln (Non-Africans Technicians and Middle Managers)	9.3E-5 (6.8E-5)			-4.7E-5 (7.2E-5)		
ln (Senegalese Upper Managers)	7.2E-5 (7.0E-5)			5.1E-5 (7.4E-5)		
ln (Other Africans Upper Managers)	2.3E-4 (1.5E-4)			1.4E-4 (1.6E-4)		
ln (Non-Africans Upper Managers)	-3.1E-5 (8.6E-5)			-7.0E-5 (9.2E-5)		
ln (Total Senegalese)			0.510* (0.057)			0.580* (0.060)
ln (Total Other Africans)			1.6E-4* (5.9E-5)			4.9E-5 (6.2E-5)

ln (Total Non-Africans)			2.9E-5 (1.1E-4)	1.2E-4$^+$ (7.5E-5)	9.2E-5 (7.1E-5)	-4.3E-5 (1.1E-4)
ln (Wage Bill Temporary Workers)	1.2E-4$^+$ (7.0E-5)	1.2E-4$^+$ (6.7E-5)	1.2E-4$^+$ (6.9E-5)			1.1E-4 (7.2E-5)
ln K	0.494* (0.043)	0.511* (0.042)	0.467* (0.046)	0.574* (0.046)	0.571* (0.045)	0.505* (0.049)
Time	-0.054* (0.016)	-0.062* (0.016)	-0.057* (0.016)	-0.076* (0.017)	-0.075* (0.016)	-0.070* (0.016)
Industry Intercepts	yes	yes	yes	yes	yes	yes
R^2	0.885	0.883	0.880	0.884	0.883	0.881
N	386	386	386	375	375	375

* = Significant at the 5 percent statistical test level.
$^+$ = Significant at the 10 percent statistical test level.
Source: 80 firm sample

to adopt more efficient methods of production, although this conclusion must be qualified by the fact that many Senegal-based firms export to other West African countries rather than to the more competitive markets in economically developed countries.

The labor input coefficients reported in columns 1 and 4 of Tables 6.1 and 6.2 suggest that output elasticities of non-Africans in all skill categories are not significantly greater than zero.[6] Put differently, one cannot reject the hypothesis that the marginal product of non-African labor in any given skill group is zero. This result indicates that non-Africans could be removed on the margin without affecting the level of output, *ceteris paribus*. Given the high salaries commanded by expatriates, this zero marginal productivity estimate suggests that from the standpoint of national economic welfare, the rapid Senegalization that we discussed in Chapter 4 could have proceeded at an even faster rate.

This conclusion has to be qualified by the fact that estimates of zero marginal product are obtained for upper managers in each one of the three nationality groups. The fully unconstrained model does not permit one to detect significant variations in productivity when the level of upper management input is varied in any national group.

To explore this issue further, we have estimated several more constrained versions of the basic model. Regressions reported in columns 3 and 6 of Tables 6.1 and 6.2 aggregate the labor input across skill categories and generate estimates of the output elasticities of the total labor input of each of the three national groups. The results based on this more parsimonious specification point to a significantly positive marginal product for the Senegalese employees as a group and an insignificantly negative marginal product for the non-Africans taken as a group. The estimate for other Africans, taken as a group, is positive and significant or insignificant, depending on specification. This more constrained specification hence also suggests that a somewhat faster exodus of non-African employees, although reducing considerably the enterprise wage bill, would have no significant effect on output.

---

6. The significant coefficient in the unskilled non-African worker category is irrelevant from the public policy standpoint because the number of these workers declined from 29 in 1980 to 14 in 1984 and to 1 in 1985.

As a further check, the IV and OLS regressions reported in Table 6.3 constrain the labor input coefficients of the three national groups to be identical, but allow for different coefficients by skill group. The results suggest that the marginal product of upper level managers, taken as a group, is significantly positive.

TABLE 6.3

Senegal (All Industries): OLS and IV Estimates of Augmented Cobb-Douglas Production Functions Based on Skill Groups
Dependent Variable = Log of Constant Price Value Added

(Values in parentheses are standard errors)

	Based on Value-Added in Domestic Prices		Based on Value-Added in World Prices	
	OLS	IV	OLS	IV
Intercept	-0.493	-0.780	-1.199*	-1.425*
% Senegalese Private	(0.489)	(0.522)	(0.530)	(0.564)
% Senegalese State	-0.074	-5.6E-4	-0.048	7.8E-3
	(0.248)	(0.268)	(0.258)	(0.277)
Ownership:				
% French	0.176+	0.179+	0.035	0.036
	(0.097)	(0.100)	(0.102)	(0.105)
% Syro-Lebanese	-0.418*	-0.370*	-0.516*	-0.471*
	(0.114)	(0.116)	(0.118)	(0.121)
% Other Foreign	0.109	0.116	-0.061	-0.054
	(0.118)	(0.121)	(0.123)	(0.124)
Trade Variables:				
Effective Rate of Protection	8.0E-4	2.0E-3	-0.325*	-0.329*
	(0.007)	(7.4E-3)	(0.021)	(0.022)
% Exports	0.026	-0.043	-0.043	-0.122
	(0.159)	(0.189)	(0.167)	(0.196)

(Continued)

Table 6.3 (Con't.)

	Based on Value-Added in Domestic Prices		Based on Value-Added in World Prices	
Inputs:	OLS	IV	OLS	IV
ln (Unskilled Workers)	0.395* (0.052)	0.368* (0.058)	0.435* (0.055)	0.421* (0.060)
ln (Skilled Workers)	1.4E-4 (1.3E-4)	2.9E-4+ (1.8E-4)	1.7E-4 (1.4E-4)	3.5E-4* (1.8E-4)
ln (Middle Management)	1.3E-4 (7.8E-5)	2.0E-4* (9.9E-5)	5.8E-5 (8.1E-5)	9.3E-5 (1.0E-4)
ln (Upper Management)	2.8E-4* (1.3E-4)	6.5E-4* (1.8E-4)	1.5E-4 (1.3E-4)	3.7E-4* (1.8E-4)
ln (Wage Bill Temporary Workers)	7.7E-5 (6.9E-5)	8.0E-5 (8.2E-5)	6.7E-5 (7.3E-5)	8.2E-5 (8.6E-5)
ln K	0.565* (0.043)	0.593* (0.047)	0.611* (0.046)	0.631* (0.050)
Time	-0.067* (0.016)	-0.070* (0.017)	-0.079* (0.017)	-0.081* (0.017)
Industry Intercept	Yes	Yes	Yes	Yes
R^2	0.873	0.869	0.874	0.872
N	386	387	375	387

* = Significant at the 5 percent statistical test level.
+ = Significant at the 10 percent statistical test level.
Source: 80-firm sample

These results are consistent with the interpretation that upper level managerial input is important, but that the observed substantial decline in the non-African managerial input has not had a significant impact on output.

Returning to Tables 6.1 and 6.2, we observe that in most specifications the wage bill of temporary workers has a significant positive coefficient. If we assume that temporary workers are paid a competitive wage, the variation in the wage bill approximates the variation in the effective use of this input. The estimates suggest that the marginal product of the temporary labor is positive and that this easily adjustable input is not used to the point where its marginal product is zero. However, as we will show, the significance of this estimated coefficient drops in the industry-specific regressions with fewer degrees of freedom.

The estimated coefficient on the time trend is negative and significant, suggesting that, *ceteris paribus*, firms have tended to experience a 5-7 percent decline in total factor productivity per year in the 1980s. While this finding is consistent with the view that the highly regulated labor market and a noncooperative system of industrial relations are having a negative productivity impact, many other factors undoubtedly contribute to this potentially serious problem.

6.2 Industry-Specific Production Function Estimates

Tables 6.4-6.13 report the IV and OLS estimates for the food industry, chemical and extraction industry, mechanical industry, textile, clothing and leather industry, and paper and wood industry. The principal advantage of these industry estimates is that they allow for inter-industry differences in technology and in the impact of various structural and policy factors. As such, they more accurately reflect the factors that determine value-added in each industry. The main drawback of estimating the regressions by industry is that the number of observations per industry and the statistical significance of the estimated coefficients are naturally smaller in industry-specific runs than in the regressions based on the entire sample.

Because the number of firms in each industry is considerably less than the total number of firms in the overall (all industry) sample, we did not rely on the canonical correlation procedure to create instrumental variables for the industry-specific regressions; rather, the instruments used in these IV regressions were firm dummy variables interacted with time and with the value-added deflator. The predictive power of these instruments was quite high and, as can be seen from Tables 6.4-6.13, the OLS estimates are only marginally more significant than their IV counterparts.

With several exceptions, the industry estimates parallel the overall (sample-wide) findings discussed in section 6.1. The closest correspondence arises with respect to the trade variables: The effective rate of protection invariably displays a negative significant effect on total factor productivity in terms of world prices, but enterprise export orientation is only rarely found to have a positive association with real value-added.

Enterprise ownership registers a somewhat more diverse impact across industries. Senegalese state and, to a lesser extent, other (than French and Syro-Lebanese) foreign ownership are found to be conducive to higher total factor productivity in the food industry; French and other foreign ownership have a similar impact in the chemical and extraction industries. Finally, there is limited evidence that other foreign firms tend to be less productive than others in the mechanical industry. In the textile, clothing and leather as well as wood and paper industries, the impact of ownership could not be precisely estimated because of collinearity problems. The serious nature of these problems is demonstrated for the textile, clothing and leather industry in columns 1 and 2 as well as 4 and 5 of Tables 6.10 and 6.11. In the case of wood and paper, we therefore present only specifications without the ownership variables.

The coefficients on unskilled or unskilled plus skilled labor are positive and significant in most regressions, but the coefficient estimates for more skilled employees, although usually positive, are invariably insignificant. As mentioned earlier, the estimated coefficients for the wage bill of temporary workers are mostly positive but they are usually insignificant.

The negative time trend effect is found in most regressions in the food and mechanical industries. However, the negative effect is less significant in the textile, clothing and leather industry and it is insignificant in the chemical and extraction industries. There is hence some evidence that the decline in total factor productivity may not be universal.

TABLE 6.4

Senegal (Food Industry): IV Estimates of Augmented Cobb-Douglas Production Functions — Dependent Variable = Log of Constant Price Value Added

(Values in parentheses are standard errors)

	Based on Value-Added in Domestic Prices			Based on Value-Added in World Prices		
	1	2	3	4	5	6
Intercept (% Senegalese Private)	0.367 (1.049)	-0.743 (0.954)	0.031 (0.912)	2.442 (1.756)	1.527 (1.791)	1.229 (1.510)
Ownership:						
% Senegalese State	1.165+ (0.682)	1.055 (0.682)	0.844 (0.593)	1.252* (0.617)	1.759* (0.530)	1.006+ (0.530)
% French	0.261 (0.330)	-0.246 (0.256)	-0.102 (0.233)	0.034 (0.388)	-0.312 (0.300)	-0.267 (0.279)
% Syro-Lebanese	0.068 (0.394)	-0.152 (0.227)	-0.202 (0.211)	0.136 (0.456)	0.160 (0.275)	0.019 (0.253)
% Other Foreign	0.661+ (0.390)	0.402 (0.275)	0.383 (0.254)	0.470 (0.461)	0.524 (0.379)	0.289 (0.336)
Trade Variables:						
Effective Rate of Protection	0.097* (0.048)	0.148* (0.044)	0.115* (0.040)	-0.353* (0.063)	-0.248* (0.064)	-0.298* (0.054)

(Continued)

Table 6.4 (Cont.)

	Based on Value-Added in Domestic Prices			Based on Value-Added in World Prices		
	1	2	3	4	5	6
% Exports	-0.320 (0.267)	-0.672* (0.282)	-0.427 (0.267)	-0.425 (0.312)	-0.725* (0.339)	-0.508 (0.326)
Inputs:						
ln (Senegalese Skilled and Unskilled)	0.651* (0.115)			0.673* (0.136)		
ln (Other African Skilled & Unskilled)	1.1E-4 (1.5E-4)			2.6E-4 (1.8E-4)		
ln (Non-African Skilled and Unskilled)	4.0E-4* (1.4E-4)			2.9E-4* (1.7E-4)		
ln (Senegalese Middle and Upper Management)	1.9E-4 (1.9E-4)			2.6E-4 (2.4E-4)		
ln (Other African Middle and Upper Management)	2.3E-4 (1.9E-4)			6.2E-4* (2.3E-4)		
ln (Non-African Middle and Upper Management)	-5.1E-4+ (3.1E-4)			-5.8E-4 (3.6E-4)		
ln (Unskilled Workers)		0.828* (0.122)	⎱ 0.657* (0.116)		0.808* (0.146)	⎱ 0.687* (0.141)
ln (Skilled Workers)		-0.001* (4.7E-4)	⎰		-9.1E-4 (5.7E-4)	⎰

ln (Middle Management)		-3.0E-4 (2.4E-4)			9.5E-6 (2.9E-4)	
ln (Upper Management)		4.1E-4 (3.6E-4)			3.4E-4 (4.6E-4)	
ln (Wage Bill of Temporary Workers)	3.8E-5 (1.8E-4)	1.9E-4 (1.6E-4)	2.4E-4 (1.5E-4)	3.1E-4 (2.4E-4)	4.8E-4* (2.2E-4)	4.7E-4* (2.0E-4)
lnK	0.351* (0.097)	0.367* (0.094)	0.369* (0.096)	0.228 (0.142)	0.201 (0.155)	0.274* (0.137)
Time	-0.072* (0.030)	-0.056* (0.034)	-0.089* (0.030)	-0.070+ (0.037)	-0.044 (0.045)	-0.088* (0.039)
R^2	0.895	0.879	0.883	0.835	0.802	0.801
N	116	116	116	110	110	110

* = Significant at the 5 % statistical test level.
\+ = Significant at the 10 % statistical test level.
The food industry includes 24 firms from the following subindustries: sugar, oil, fish, flour, and other foods.

TABLE 6.5

Senegal (Food Industry): OLS Estimates of Augmented Cobb-Douglas Production Functions—Dependent Variable = Log of Constant Price Value Added

(Values in parentheses are standard errors)

	Based on Value-Added in Domestic Prices			Based on Value-Added in World Prices		
	1	2	3	4	5	6
Intercept (% Senegalese Private)	0.394 (0.845)	-0.623 (0.711)	0.033 (0.742)	2.665+ (1.546)	1.626 (1.531)	1.312 (1.381)
Ownership:						
% Senegalese State	1.255* (0.455)	1.164* (0.386)	0.859* (0.368)	1.473* (0.581)	1.710* (0.569)	1.082* (0.513)
% French	0.276 (0.319)	-0.124 (0.237)	-0.069 (0.228)	0.079 (0.374)	-0.232 (0.287)	-0.227 (0.275)
% Syro-Lebanese	0.107 (0.374)	-0.189 (0.209)	-0.199 (0.207)	0.248 (0.436)	0.122 (0.264)	0.034 (0.253)
% Other Foreign	0.703+ (0.364)	0.466* (0.233)	0.412+ (0.231)	0.601 (0.439)	0.581+ (0.348)	0.351 (0.326)
Trade Variables:						
Effective Rate of Protection	0.098* (0.044)	0.118* (0.039)	0.109* (0.038)	-0.322* (0.058)	-0.263* (0.059)	-0.295* (0.054)

	(1)	(2)	(3)	(4)	(5)	(6)
% Exports	-0.342 (0.227)	-0.391+ (0.231)	-0.383+ (0.229)	-0.451+ (0.266)	-0.446 (0.284)	-0.435 (0.282)
Inputs:						
ln (Senegalese Skilled and Unskilled)	0.632* (0.098)			0.663* (0.115)		
ln (Other African Skilled and Unskilled)	0.11E-3 (0.13E-3)			0.27E-3+ (0.16E-3)		
ln (Non-African Skilled and Unskilled)	0.28E-3* (0.11E-3)			0.20E-3 (0.13E-3)		
ln (Senegalese Middle and Upper Management)	0.20E-3 (0.18E-3)			0.32E-3 (0.23E-3)		
ln (Other African Middle and Upper Management)	0.16E-3 (0.16E-3)			0.43E-3* (0.19E-3)		
ln (Non-African Middle and Upper Management)	-0.49E-3+ (0.29E-3)			-0.59E-3+ (0.35E-3)		
ln (Unskilled Workers)		0.638* (0.097)	⎫ 0.632* ⎬ (0.101)		0.651* (0.119)	⎫ 0.660* ⎬ (0.122)
ln (Skilled Workers)		-0.39E-3 (0.33E-3)	⎭		-0.37E-3 (0.41E-3)	⎭

(Continued)

Table 6.5 (Con't.)

	Based on Value-Added in Domestic Prices			Based on Value-Added in World Prices		
	1	2	3	4	5	6
ln (Middle Management)		-0.22E-3 (0.21E-3)			0.30E-4 (0.26E-3)	
ln (Upper Management)		0.13E-3 (0.31E-3)			0.54E-4 (0.39E-3)	
ln (Wage Bill of Temporary Workers)	0.14E-3 (0.15E-3)	0.20E-3 (0.14E-3)	0.24E-3$^+$ (0.14E-3)	0.39E-3* (0.19E-3)	0.46E-3* (0.19E-3)	0.45E-3* (0.18E-3)
lnK	0.349* (0.078)	0.429* (0.072)	0.377* (0.079)	0.194 (0.126)	0.251$^+$ (0.135)	0.274* (0.125)
Time	-0.076* (0.029)	-0.081* (0.031)	-0.090* (0.029)	-0.070* (0.036)	-0.068$^+$ (0.041)	-0.089* (0.038)
R^2	0.897	0.888	0.883	0.838	0.808	0.801
N	115	115	115	109	109	109

* = Significant at the 5 % statistical test level.
+ = Significant at the 10 % statistical test level.
The food industry includes 24 firms from the following sub-industries: sugar, oil, fish, flour, and other food.

TABLE 6.6

Senegal (Chemical & Extraction Industries):
IV Estimates of Augmented Cobb-Douglas Production
Functions — Dependent Variable = Log of Constant Price Value Added

(Values in parentheses are standard errors)

	Based on Value-Added in Domestic Price			Based on Value-Added in World Prices		
	1	2	3	4	5	6
Intercept (% Senegalese Private)	-2.117 (2.735)	-3.253* (1.422)	-1.951 (1.567)	-3.651+ (2.083)	-3.799* (1.379)	-2.644+ (1.465)
Ownership:						
% Senegalese State	-0.703 (2.239)	-0.564 (1.369)	-1.039 (1.280)	-2.353 (1.657)	-1.079 (1.067)	-0.813 (0.947)
% French	1.356* (0.568)	0.992* (0.370)	0.951* (0.371)	0.134 (0.403)	0.639* (0.296)	0.427 (0.272)
% Syro-Lebanese	0.625 (0.493)	0.461 (0.367)	0.266 (0.347)	0.384 (0.384)	0.426 (0.345)	0.243 (0.291)
% Other Foreign	2.259* (0.702)	1.402* (0.452)	1.319* (0.438)	0.929 (0.649)	0.816+ (0.422)	0.633+ (0.354)
Trade Variables:						
Effective Rate of Protection	0.052+ (0.029)	0.032+ (0.019)	0.021 (0.022)	-0.328* (0.065)	-0.341* (0.054)	-0.358* (0.049)

Table 6.6 (Con't)

	Based on Value-Added in Domestic Price			Based on Value-Added in World Prices		
	1	2	3	4	5	6
% Exports	1.978 (1.350)	1.541⁺ (0.933)	1.964* (0.853)	1.613* (0.792)	1.179* (0.603)	1.076* (0.600)
Inputs:						
ln (Senegalese Skilled and Unskilled)	-0.050 (0.285)			0.388 (0.262)		
ln (Other African Skilled and Unskilled)	-1.2E-4 (3.0E-4)			-6.2E-4* (2.3E-4)		
ln (Non-African Skilled and Unskilled)	1.2E-4 (3.0E-4)			9.9E-5 (2.3E-4)		
ln (Senegalese Middle and Upper Management)	0.001* (5.4E-4)			3.2E-4 (5.9E-4)		
ln (Other African Middle and Upper Management)	2.6E-4 (3.2E-4)			2.3E-4 (2.2E-4)		
ln (Non-African Middle and Upper Management)	3.5E-4 (6.0E-4)			-5.4E-5 (5.1E-4)		

	(1)	(2)	(3)	(4)	(5)	(6)
ln(Unskilled Workers)		-0.037 (0.160)	0.244 (0.174)		0.111 (0.159)	0.251 (0.160)
ln(Skilled Workers)		6.8E-4 (1.2E-3)	⎫		-6.9E-4 (1.2E-3)	⎫
ln(Middle Management)		4.2E-4 (3.1E-4)	⎬		-4.5E-5 (3.2E-4)	⎬
ln(Upper Management)		1.1E-4 (4.7E-4)	⎭		2.3E-4 (4.6E-4)	⎭
ln(Wage Bill of Temporary Workers)	-5.3E-4 (4.7E-4)	-2.4E-4 (3.0E-4)	-1.2E-4 (3.0E-4)	5.3E-4 (3.5E-4)	2.7E-4 (2.6E-4)	2.7E-4 (2.3E-4)
lnK	0.655* (0.250)	0.748* (0.145)	0.545* (0.161)	0.706* (0.202)	0.787* (0.136)	0.659* (0.150)
Time	-0.023 (0.055)	-0.031 (0.038)	-0.004 (0.038)	-0.054 (0.043)	-0.036 (0.038)	-0.025 (0.035)
R^2	0.805	0.867	0.857	0.891	0.885	0.889
	92	92	92	87	87	87

* = Significant at the 5 % statistical test level.
+ = Significant at the 10 % statistical test level.
Our sample in these industries consists of 18 firms.

TABLE 6.7

**Senegal (Chemical and Extraction Industries):
OLS Estimates of Augmented Cobb-Douglas Production
Functions — Dependent Variable = Log of Constant Price Value Added**

(Values in parentheses are standard errors)

	Based on Value-Added in Domestic Prices			Based on Value-Added in World Prices		
	1	2	3	4	5	6
Intercept	-2.506+ (1.057)	-3.281* (1.162)	-2.071+ (1.609)	-3.871* (1.204)	-3.749* (1.306)	-2.939* (1.349)
% Senegalese Private)						
Ownership:						
% Senegalese State	1.514+ (0.829)	1.086 (0.738)	1.138 (0.718)	-0.485 (0.957)	0.027 (0.836)	0.168 (0.805)
% French	0.779* (0.239)	0.793* (0.221)	0.690* (0.217)	0.411 (0.286)	0.656* (0.248)	0.549* (0.242)
% Syro-Lebanese	0.232 (0.268)	0.295 (0.266)	0.050 (0.244)	0.397 (0.305)	0.403 (0.295)	0.309 (0.271)
% Other Foreign	1.498* (0.359)	1.133* (0.306)	0.939* (0.289)	0.973* (0.485)	0.716* (0.367)	0.644+ (0.333)
Trade Variables:						
Effective Rate of Protection	0.013 (0.012)	0.013 (0.011)	0.001 (0.011)	-0.344* (0.054)	-0.331* (0.049)	-0.343* (0.047)

	(1)	(2)	(3)	(4)	(5)	(6)
% Exports	0.257 (0.387)	0.405 (0.389)	0.452 (0.372)	0.649 (0.447)	0.567 (0.439)	0.545 (0.415)
Inputs:						
ln (Senegalese Skilled and Unskilled)	0.109 (0.143)			0.205 (0.170)		
ln (Other African Skilled and Unskilled)	-0.11E-3 (0.15E-3)			-0.34E-3* (0.17E-3)		
ln (Non-African Skilled and Unskilled)	-0.01E-3 (0.15E-3)			-0.33E-4 (0.19E-3)		
ln (Senegalese Middle and Upper Management)	0.88E-3* (0.29E-3)			0.42E-3 (0.40E-3)		
ln (Other African Middle and Upper Management)	0.14E-3 (0.14E-3)			0.22E-3 (0.15E-3)		
ln (Non-African Middle and Upper Management)	0.29E-3 (0.25E-3)			0.27E-3 (0.28E-3)		
ln (Unskilled Workers)		0.029 (0.113) ⎫	0.291* (0.125)		0.088 (0.127) ⎫	0.210 (0.141)
ln (Skilled Workers)		0.17E-3 (0.51E-3) ⎬			-0.17E-4 (0.57E-3) ⎬	
ln (Middle Management)		0.37E-3+ (0.20E-3) ⎭			-0.85E-5 (0.24E-3) ⎭	

(Continued)

Table 6.7 (Con't.)

	Based on Value-Add in Domestic Prices			Based on Value-Added in World Prices		
	1	2	3	4	5	6
ln (Upper Management)		-0.82E-4 (0.35E-3)			0.12E-3 (0.38E-3)	
ln (Wage Bill of Temporary Workers)	-0.11E-3 (0.15E-3)	-0.94E-4 (0.14E-3)	-0.34E-4 (0.14E-3)	0.82E-4 (0.17E-3)	-0.16E-4 (0.17E-3)	0.89E-5 (0.16E-3)
lnK	0.661* (0.126)	0.748* (0.101)	0.563* (0.118)	0.773* (0.151)	0.789* (0.115)	0.688* (0.132)
Time	-0.016 (0.032)	-0.031 (0.032)	-0.59E-3 (0.031)	-0.049 (0.036)	-0.037 (0.035)	-0.029 (0.034)
R^2	0.915	0.900	0.900	0.907	0.892	0.894
N	91	91	91	86	86	86

* = Significant at the 5 % statistical test level.
+ = Significant at the 10 % statistical test level.
Our sample in these industries consists of 18 firms.

TABLE 6.8
Senegal (Mechanical Industry):
IV Estimates of Augmented Cobb-Douglas Production
Functions — Dependent Variable = Log of Constant Price Value Added

(Values in parentheses are standard errors)

	Based on Value-Added in Domestic Prices			Based on Value-Added in World Prices		
	1	2	3	4	5	6
Intercept (% Senegalese Private)	2.732 (1.857)	-0.065 (1.094)	-0.845 (0.699)	1.863 (2.034)	0.050 (1.251)	-1.660* (0.797)
Ownership:						
% Senegalese State	1.406 (1.483)	-0.059 (1.301)	-0.520 (1.024)	2.077 (1.625)	1.049 (1.487)	0.648 (1.168)
% French	0.216 (0.388)	0.207 (0.503)	0.285 (0.270)	-0.494 (0.425)	-0.443 (0.575)	-0.230 (0.308)
% Syro-Lebanese	1.466 (0.913)	1.213 (1.332)	-0.429 (0.510)	1.985* (1.000)	2.004 (1.522)	0.016 (0.581)
% Other Foreign	-0.441 (0.348)	-0.533 (0.480)	-0.574+ (0.292)	-1.151* (0.382)	-1.137* (0.549)	-1.248* (0.333)
Trade Variables:						
Effective Rate of Protection	0.037 (0.037)	0.017 (0.032)	0.027 (0.027)	-0.239* (0.040)	-0.281* (0.037)	-0.258* (0.031)

(Continued)

Table 6.8 (Con't.)

	Based on Value-Added in Domestic Prices			Based on Value-Added in World Prices		
	1	2	3	4	5	6
% Exports	-0.025 (0.853)	1.252[+] (0.746)	0.370 (0.621)	-0.110 (0.934)	1.626[+] (0.853)	0.208 (0.707)
Inputs						
ln (Senegalese Skilled and Unskilled)	0.476 (0.386)			0.418 (0.422)		
ln (Other African Skilled and Unskilled)	1.7E-4 (3.4E-4)			4.4E-4 (3.8E-4)		
ln (Non-African Skilled and Unskilled)	8.0E-4[*] 3.9E-4			7.6E-4[+] (4.3E-4)		
ln (Senegalese Middle and Upper Management)	-1.6E-4 (3.1E-4)			-1.9E-4 (3.4E-4)		
ln (Other African Middle and Upper Management)	7.8E-4 (5.5E-4)			9.8E-4 (6.1E-4)		
ln (Non-African Middle and Upper Management)	-3.0E-4 (6.0E-4)			-4.4E-4 (6.8E-4)		
ln (Unskilled)		0.656[*] (0.215)	0.732[*] (0.176)		0.682[*] (0.246)	0.813[*] (0.200)
ln (Skilled)		6.5E-4 (4.2E-4)			9.6E-4[*] (4.8E-4)	

	(1)	(2)	(3)	(4)	(5)	(6)
ln (Middle Management)		1.8E-4 (3.7E-4)			4.4E-4 (4.3E-4)	-3.4E-4 (2.7E-4)
ln (Upper Management)		7.8E-4* (3.9E-4)			8.9E-4* (4.4E-4)	
ln (Wage Bill Temporary Workers)	1.9E-4 (3.9E-4)	-7.2E-5 (2.7E-4)	-1.4E-4 (2.3E-4)	1.9E-4 (4.3E-4)	-1.9E-4 (3.1E-4)	
lnK	0.272 (0.204)	0.383* (0.112)	0.394* (0.092)	0.417$^+$ (0.224)	0.394* (0.128)	0.451* (0.105)
Time	0.035 (0.060)	-0.108* (0.053)	-0.070$^+$ (0.038)	0.036 (0.065)	-0.122* (0.060)	-0.097* (0.043)
R^2	0.913	0.896	0.913	0.929	0.907	0.923
N	54	54	54	54	54	54

* = Significant at the 5 % statistical test level.
$^+$ = Significant at the 10 % statistical test level.
Our sample in this industry consists of 13 firms.

TABLE 6.9
Senegal (Mechanical Industry):
OLS Estimates of Augmented Cobb-Douglas Production
Functions — Dependent Variable = Log of Constant Price Value Added

(Values in parentheses are standard errors)

	Based on Value-Added in Domestic Prices			Based on Value-Added in World Prices		
	1	2	3	4	5	6
Intercept (% Senegalese Private)	1.396 (1.248)	-0.864 (0.884)	-0.907 (0.683)	0.449 (1.365)	-1.109 (0.985)	-1.711* (0.768)
Ownership:						
% Senegalese State	0.576 (1.026)	-0.477 (0.916)	-0.843 (0.759)	1.116 (1.122)	0.599 (1.019)	-0.038 (0.854)
% French	0.297 (0.299)	0.197 (0.357)	0.309 (0.259)	-0.395 (0.328)	-0.417 (0.398)	-0.227 (0.291)
% Syro-Lebanese	0.481 (0.686)	0.050 (0.829)	-0.481 (0.493)	0.893 (0.750)	0.601 (0.923)	-0.092 (0.554)
% Other Foreign	-0.515+ (0.301)	-0.634+ (0.347)	-0.494+ (0.282)	-1.231* (0.329)	-1.263* (0.386)	-1.139* (0.317)
Trade Variables:						
Effective Rate of Protection	0.017 (0.030)	0.018 (0.029)	0.026 (0.026)	-0.254* (0.033)	-0.278* (0.033)	-0.257* (0.030)

	(1)	(2)	(3)	(4)	(5)	(6)
% Exports	0.196 (0.370)	0.735+ (0.379)	0.488 (0.350)	0.202 (0.405)	0.856* (0.423)	0.506 (0.394)
Inputs:						
ln (Senegalese Skilled and Unskilled)	0.468* (0.172)			0.442* (0.188)		
ln (Other African Skilled and Unskilled)	0.89E-4 (0.18E-3)			0.31E-3 (0.20E-3)		
ln (Non-African Skilled and Unskilled)	0.39E-3+ (0.23E-3)			0.32E-3 (0.25E-3)		
ln (Senegalese Middle and Upper Management)	0.53E-5 (0.19E-3)			-0.33E-4 (0.21E-3)		
ln (Other African Middle and Upper Management)	0.55E-3* (0.27E-3)			0.69E-3* (0.29E-3)		
ln (Non-African Middle and Upper Management)	0.27E-3 (0.38E-3)			0.19E-3 (0.42E-3)		
ln (Unskilled Workers)		0.632* (0.157)	⎫ 0.659* ⎬ (0.136)		0.673* (0.174)	⎫ 0.703* ⎬ (0.153)
ln (Skilled Workers)		0.24E-3 (0.22E-3)	⎭		0.43E-3+ (0.24E-3)	⎭

(Continued)

Table 6.9 (Con't.)

	Based on Value-Added in Domestic Prices			Based on Value-Added in World Prices		
	1	2	3	4	5	6
ln (Middle Management)		0.15E-3 (0.24E-3)			0.34E-3 (0.27E-3)	
ln (Upper Management)		0.42E-3 (0.28E-3)			0.45E-3 (0.32E-3)	
ln (Wage Bill of Temporary Workers)	0.45E-4 (0.19E-3)	-0.99E-4 (0.17E-3)	-0.41E-4 (0.15E-3)	0.28E-4 (0.20E-3)	-0.23E-3 (0.19E-3)	-0.14E-3 (0.17E-3)
lnK	0.361* (0.119)	0.457* (0.096)	0.424* (0.084)	0.491* (0.131)	0.489* (0.107)	0.496* (0.095)
Time	-0.024 (0.043)	-0.091* (0.041)	-0.063+ (0.034)	-0.029 (0.047)	-0.106* (0.046)	-0.083* (0.038)
R^2	0.926	0.910	0.914	0.939	0.924	0.926
N	53	53	53	53	53	53

* = Significant at the 5 % statistical test level.
+ = Significant at the 10 % statistical test level.
Our sample in this industry consists of 13 firms.

TABLE 6.10

Senegal (Textile, Clothing and Leather Industries: IV Estimates of Augmented Cobb-Douglas Production Functions — Dependent Variable = Log of Constant Price Value Added

(Values in parentheses are standard errors)

	Based on Value-Added in Domestic Prices			Based on Value-Added in World Prices		
	1	2	3	4	5	6
Intercept (% Senegalese Private)	-0.604 (6.766)	18.389* (6.977)	0.451 (1,789)	-0.766 (6.787)	18.830* (6.970)	0.268 (1.800)
Ownership:						
% Senegalese State	-1.342+ (0.777)	2.133* (0.883)		-1.309+ (0.780)	2.161* (0.883)	
% French	-0.423 (1.387)	-3.038* (1.336)		-0.398 (1.391)	-3.123* (1.335)	
% Syro-Lebanese	-0.612 (0.814)	-2.459* (0.790)		-0.651 (0.817)	-2.484* (0.789)	
% Other Foreign	-2.279 (5.361)	2.658 (8.455)		-2.629 (5.378)	2.902 (8.449)	
Trade Variables:						
Effective Rate of Protection	0.401 (0.764)	-0.780 (1.063)	-0.232 (0.208)	-0.183 (0.767)	-1.440 (1.062)	-0.863* (0.209)

(Continued)

Table 6.10 (Con't.)

	Based on Value-Added in Domestic Prices			Based on Value-Added in World Prices		
	1	2	3	4	5	6
% Exports	-1.492 (1.291)	-1.284 (1.159)	0.171 (0.316)	-1.528 (1.295)	-1.233 (1.159)	0.192 (0.318)
Inputs:						
ln (Senegalese Skilled and Unskilled)	1.272+ (0.646)			1.262+ (0.648)		
ln (Other African Skilled and Unskilled)	-3.9E-4 (6.3E-4)			-3.6E-4 (6.3E-4)		
ln (Non-African Skilled and Unskilled)	-3.2E-4 (5.2E-4)			-3.4E-4 (5.2E-4)		
ln (Senegalese Middle and Upper Management)	-7.2E-5 (5.4E-4)			-9.8E-5 (5.4E-4)		
ln (Other African Middle and Upper Management)	5.3E-5 (4.5E-4)			4.2E-5 (4.5E-4)		
ln (Non-African Middle and Upper Management)	-0.002* (4.3E-4)			-0.002* (4.4E-4)		
ln (Unskilled Workers)		0.365 (0.423)	0.906* (0.360)		0.382 (0.422)	0.849* (0.362)
ln (Skilled Workers)		-4.6E-4 (8.4E-4)			-4.7E-4 (8.4E-4)	

	(1)	(2)	(3)	(4)	(5)	(6)
ln (Middle Management)		0.002* (5.5E-4)			0.002* (5.5E-4)	
ln (Upper Management)		0.794 (0.635)			0.750 (0.635)	
ln (Wage Bill of Temporary Workers)	6.2E-4 (4.2E-4)	4.6E-4 (3.9E-4)	1.3E-4 (2.0E-4)	6.2E-4 (4.2E-4)	4.6E-4 (3.9E-4)	1.1E-4 (2.0E-4)
lnK	0.152 (0.613)	-0.761 (0.486)	0.204 (0.266)	0.164 (0.614)	-0.795 (0.486)	0.239 (0.268)
Time	-0.074 (0.069)	-0.051 (0.045)	-0.053 (0.033)	-0.076 (0.070)	-0.049 (0.045)	-0.056[+] (0.033)
R^2	0.946	0.923	0.923	0.962	0.945	0.939
N	51	51	62	51	51	62

* = Significant at the 5 % statistical test level.
+ = Significant at the 10 % statistical test level.
Our sample in these industries consists of 11 firms.

TABLE 6.11
Senegal (Textile, Clothing and Leather Industries)
OLS Estimates of Augmented Cobb-Douglas Production
Functions — Dependent Variable = Log of Constant Price Value Added

(Values in parentheses are standard errors)

	Based on Value-Added in Domestic Prices			Based on Value-Added in World Prices		
	1	2	3	4	5	6
Intercept (% Senegalese Private)	-1.269 (3.480)	8.231+ (4.462)	0.227 (1.287)	-1.213 (3.488)	8.595+ (4.468)	-0.083 (1.292)
Ownership:						
% Senegalese State	-0.876+ (0.453)	0.682 (0.575)		-0.851+ (0.454)	0.706 (0.576)	
% French	0.569 (0.552)	-1.261 (0.822)		0.565 (0.553)	-1.315 (0.823)	
% Syro-Lebanese	0.204 (0.349)	-1.089* (0.486)		0.160 (0.349)	-1.128* (0.487)	
% Other Foreign	-5.298* (2.180)	-0.341 (4.517)		-5.587* (2.185)	-0.383 (4.524)	
Trade Variables:						
Effective Rate of Protection	0.318 (0.335)	-0.504 (0.560)	-0.235* (0.103)	-0.282 (0.336)	-1.129* (0.561)	-0.881* (0.103)

	(1)	(2)	(3)	(4)	(5)	(6)
% Exports	-0.344 (0.583)	-0.203 (0.636)	0.125 (0.276)	-0.347 (0.584)	-0.175 (0.636)	0.134 (0.277)
Inputs:						
ln (Senegalese Skilled and Unskilled)	1.363* (0.363)			1.357* (0.364)		
ln (Other African Skilled and Unskilled)	-0.51E-5 (0.23E-3)			0.10E-4 (0.23E-3)		
ln(Non-African Skilled and Unskilled)	0.19E-3 (0.21E-3)			0.17E-3 (0.21E-3)		
ln (Senegalese Middle and Upper Management)	0.49E-4 (0.27E-3)			0.32E-4 (0.27E-3)		
ln (Other African Middle and Upper Management)	0.33E-3+ (0.18E-3)			0.33E-3+ (0.18E-3)		
ln (Non-African Middle and Upper Management)	-0.13E-2* (0.26E-3)			-0.14E-2* (0.26E-3)		
ln (Unskilled Workers)		0.431 (0.309)	0.867* (0.160)		0.449 (0.309)	0.766* (0.160)
ln (Skilled Workers)		-0.33E-3 (0.46E-3)			-0.34E-3 (0.46E-3)	
ln (Middle Management)		0.11E-2* (0.34E-3)			0.11E-2* (0.34E-3)	

(Continued)

Table 6.11 (Con't.)

	Based on Value-Added in Domestic Prices			Based on Value-Added in World Prices		
	1	2	3	4	5	6
ln (Upper Management)		0.571$^+$ (0.340)			0.550 (0.341)	
ln (Wage Bill of Temporary Workers)	0.25E-3 (0.20E-3)	0.25E-3 (0.22E-3)	1.3E-4 (1.3E-3)	0.25E-3 (0.20E-3)	0.26E-3 (0.22E-3)	1.0E-4 (1.3E-4)
lnK	0.153 (0.346)	-0.134 (0.319)	0.238 (0.149)	0.149 (0.347)	-0.167 (0.319)	0.296* (0.150)
Time	-0.034 (0.039)	-0.068$^+$ (0.037)	-0.054$^+$ (0.032)	-0.034 (0.039)	-0.065$^+$ (0.037)	-0.057$^+$ (0.032)
R^2	0.963	0.944	0.923	0.974	0.960	0.940
N	50	50	61	50	50	61

* = Significant at the 5 % statistical test level.
+ = Significant at the 10 % statistical test level.
Because of a relatively serious multicollinearity problem, columns 3 and 6 present estimates based on a specification without ownership variables. Our sample in these industries consists of 11 firms.

TABLE 6.12
Senegal (Paper & Wood Industry):
IV Estimates of Augmented Cobb-Douglas Production
Functions — Dependent Variable = Log of Constant Price Value Added

(Values in parentheses are standard errors)

	Based on Value-Added in Domestic Prices			Based on Value-Added in World Prices		
	1	2	3	4	5	6
Intercept	4.396 (2.712)	6.789 (12.908)	2.392* (0.607)	3.588 (3.571)	8.093 (15.066)	2.600* (0.642)
Trade Variables:						
Effective Rate of Protection	-0.149 (0.235)	-0.192 (0.165)	-0.188* (0.060)	-0.533+ (0.309)	-0.696* (0.192)	-0.679* (0.063)
% Exports	-2.012 (1.725)	0.239 (4.272)	-0.451 (0.677)	-3.016 (2.270)	0.058 (4.986)	-0.629 (0.716)
Inputs:						
ln (Senegalese Skilled and Unskilled)	0.786 (0.588)			0.525 (0.774)		
ln (Other African Skilled and Unskilled)	-9.5E-4 (8.7E-4)			-0.001 (0.001)		

(Continued)

Table 6.12 (Con't.)

	Based on Value-Added in Domestic Prices			Based on Value-Added in World Prices		
	1	2	3	4	5	6
ln (Non-African Skilled and Unskilled)	0.001 (0.001)			1.9E-3 (1.9E-3)		
ln (Senegalese Middle and Upper Management)	-8.5E-5 (5.9E-4)			-3.7E-4 (7.8E-4)		
ln (Other African Middle and Upper Management)	0.001 (9.6E-4)			1.0E-3 (1.3E-3)		
ln (Unskilled Workers)		0.848 (0.637)	0.840* (0.143)		0.808 (0.743)	0.832* (0.152)
ln (Skilled Workers)		7.3E-4 (1.1E-3)			1.1E-3 (1.2E-3)	
ln (Middle Management)		-9.4E-5 (8.7E-4)			-1.1E-4 (1.0E-3)	
ln (Upper Management)		2.7E-3 (6.2E-3)			2.5E-3 (7.2E-3)	
ln (Wage Bill Temporary Workers)	2.8E-4 (4.3E-4)	1.1E-4 (9.2E-4)	3.2E-4 (2.3E-4)	6.5E-4 (5.7E-4)	1.8E-4 (1.1E-3)	5.5E-4* (2.4E-4)

lnK	0.029 (0.388)	-0.203 (1.181)	0.145* (0.066)	0.208 (0.510)	-0.283 (1.378)	0.142* (0.070)
Time	-0.056 (0.051)	-0.048 (0.103)	-0.064+ (0.035)	-0.073 (0.067)	-0.054 (0.120)	-0.073+ (0.037)
R^2	0.796	0.185	0.853	0.827	0.456	0.919
N	59	59	59	59	59	59

* = Significant at the 5 % statistical test level.
\+ = Significant at the 10 % statistical test level.
Our sample in this industry consists of 10 and it contains no state-owned firms or firms employing other African middle or upper managers. Many firms also display missing values for ownership and in the remaining firms the ownership variables are highly correlated with the other explanatory variables (especially the effective rate of protection). The ownership variables are hence excluded in the above regressions.

TABLE 6.13

Senegal (Paper and Wood Industry): OLS Estimates of Augmented Cobb-Douglas Production Functions – Dependent Variable = Log of Constant Price Value Added

(Values in parentheses are standard errors)

	Based on Value-Added in Domestic Prices			Based on Value-Added in World Prices		
	1	2	3	4	5	6
Intercept	1.553* (0.761)	1.156+ (0.671)	1.872* (0.441)	1.871* (0.805)	1.459* (0.704)	2.092* (0.468)
Trade Variables:						
Effective Rate of Protection	-0.061 (0.062)	-0.122* (0.052)	-0.126* (0.049)	-0.541* (0.066)	-0.615* (0.055)	-0.611* (0.052)
% Exports	-0.712 (0.567)	-0.026 (0.516)	0.123 (0.498)	-1.040+ (0.500)	-0.297 (0.541)	-0.103 (0.529)
Inputs:						
ln (Senegalese Skilled and Unskilled)	0.485* (0.139)			0.487* (0.147)		
ln (Other African Skilled and Unskilled)	-0.26E-3+ (0.13E-3)			-0.29E-3* (0.14E-3)		

	(1)	(2)	(3)	(4)
ln (Non-African Skilled and Unskilled)	0.36E-3+ (0.21E-3)		0.38E-3+ (0.23E-3)	
ln (Senegalese Middle and Upper Management)	0.36E-4 (0.17E-3)		0.74E-4 (0.18E-3)	
ln (Non-African Middle Upper Management)	-0.31E-3 (0.28E-3)		-0.28E-3 (0.30E-3)	
ln (Unskilled Workers)		0.377* (0.111) ⎱ 0.507* (0.104)		0.339* (0.117) ⎱ 0.504* (0.111)
ln (Skilled Workers)		0.20E-3 (0.20E-3) ⎰		0.39E-3+ (0.21E-3) ⎰
ln (Middle Management)		-0.27E-3 (0.17E-3)		-0.28E-3 (0.18E-3)
ln (Upper Management)		-0.27E-3+ (0.15E-3)		-0.25E-3 (0.16E-3)
ln (Wage Bill of Temporary Workers)	0.26E-3 (0.16E-3)	0.23E-3 (0.16E-3) 0.35E-3* (0.14E-3)	0.38E-3* (0.17E-3)	-0.33E-3* (0.17E-3) 0.47E-3* (0.15E-3)

(Continued)

Table 6.13 (Con't.)

	Based on Value-Added in Domestic Prices			Based on Value-Added in World Prices		
	1	2	3	4	5	6
lnK	0.331* (0.066)	0.380* (0.067)	0.284* (0.043)	0.313* (0.070)	0.377* (0.070)	0.274* (0.046)
Time	-0.082* (0.032)	-0.065* (0.031)	-0.062* (0.031)	-0.090* (0.033)	-0.073* (0.032)	-0.067* (0.033)
R^2	0.903	0.894	0.882	0.947	0.943	0.935
N	58	58	58	58	58	58

* = Significant at the 5 % statistical test level or better.
+ = Significant at the 10 % statistical test level or better.
Our sample in this industry consists of 10 firms and it contains no state-owned firms or firms employing other African middle or upper managers. Many firms also display missing values for ownership and in the remaining firms the ownership variables are highly correlated with the other explanatory variables (especially the effective rate of protection). The ownership variables are hence excluded in the above regressions.

7

Econometric Estimates of the Determinants of Wages

In this section we use individual worker data from the 17-firm sample to explain wage differentials in terms of productivity-related and other characteristics of individuals. Because the cross tabulations in Chapter 4 indicate that earnings differentials widen when viewed in terms of an individual's total compensation as opposed to base salary, we estimate the earnings functions using both total compensation and base salary as the dependent variables.

The results from this exercise are presented in Table 7.1. They indicate that the male-female earnings differential is significant and widens when total compensation is considered, in comparison to the base salary. Controlling for differences in education, experience, nationality and other explanatory variables, males earn on average 17.8 percent more than females in base salary and 19.2 percent in total compensation.

Earnings differentials are enormous between the non-African and African groups: After controlling for differences in education, experience, gender, and other explanatory variables, non-Africans are on average found to earn from 123 percent to 142 percent more than Africans.[1] In contrast, there is no noticeable earnings differential between similar other Africans and Senegalese, which is consistent with the information gathered from our interviews with the managers. Other Africans tend to be employed in the

---

1. The differential between non-Africans and Africans does not increase when total compensation is considered. This may be accurate as we understand that the extra pay of expatriates comes in the form of both extra base pay and supplements. Unfortunately, some of the extra supplements come in benefits "in kind," such as housing, for which we could not gather data and which would probably increase the relative total earnings differential.

lower skilled jobs and employers find no need to pay them more than comparable Senegalese.

TABLE 7.1

Senegal: Estimates of Individual Earnings Functions Based on Personal and Enterprise Characteristics

(Values in parentheses are standard errors)

	Dependent Variable = Monthly Base Salary 1	Dependent Variable = Monthly Total Earnings 2
Intercept	9.639* (0.137)	9.942* (0.194)
Male	0.178* (0.071)	0.192** (0.100)
Other African	0.255 (0.188)	.237 (0.266)
Non-African	1.420* (0.193)	1.235* (0.273)
Union Member	-0.016 (0.060)	-0.133 (0.085)
Education	0.074* (0.006)	0.111* (0.007)
Labor Force Experience	0.007* (0.003)	0.016* (0.004)
Firm-specific Experience	0.018* (0.003)	0.020* (0.004)
Training	0.108* (0.038)	0.098** (0.054)
% Senegalese State Owned	-0.024 (0.273)	0.669** (0.385)

% French	0.284* (0.083)	0.117 (0.116)
% Syro-Lebanese	-0.084 (0.090)	-0.293* (0.125)
% Other Foreign	(0.306)* (0.118)	0.748* (0.168)
Effective Rate of Protection	0.020 (0.012)	-0.005 (0.017)
% Exports	0.217* (0.083)	0.137 (0.118)
Net Revenue Per Worker	-0.30E-4 (0.23E-4)	-0.29E-4 (0.33E-4)
Sales Growth Over 5 Years	0.820* (0.323)	1.018* (0.456)
R^2	0.45	0.49
N	513	513

* = significant at 5 % statistical test level.
\+ = significant at 10 % statistical test level.
Source: Worker data from 17 firms.

The results in Table 7.1 suggest that union membership does not significantly improve most workers' base wage or total earnings. Once the effects of firm ownership and the worker's nationality, sex, education, and experience are taken into account, trade union membership does not appear to add to one's earnings. This finding could be explained by the fact that our sample contains very few nonunionized workers and that the government extends the terms of the collective agreement to nonunionized workers as well. However, these findings are also consistent with those from our interviews, where it became clear that unions are seen primarily as protectors of the workers' job rights and enforcers of the Labor Code. For example, they protect the day laborer's right to obtain a permanent job after having worked for a specified length of time.

Each additional year of education raises the base salary by 7.4 percent and the total compensation by 11.1 percent. These estimates fall within the range of the estimates of the effects of education on earnings in the Third World.[2] Moreover, the higher return in total compensation (as compared to base salary) reflects the extra premiums that are given to the more educated middle level managers.

We looked at experience in terms of work prior to and within the firm of current employment to test the hypothesis that there is a higher return on each year of experience within the firm of employment than on years of experience prior to entering the firm. The results indicate that the return is indeed higher for firm-specific experience (1.8 percent) compared to general labor force experience (0.7 percent) for one's base salary. However, one year of experience within or outside the firm has approximately the same payoff for total earnings--2 percent vs. 1.6 percent, respectively. Managers hence appear to be awarding greater supplements to employees with more labor force experience. This is also clear from comparing the coefficient on labor force experience in columns 1 and 2 of Table 7.1.

The coefficient on training indicates that those who have received some training in the firm earn approximately 10 percent more each month than those who have not been trained.[3] Is this earnings differential the result of internal training--that is, does training bring about a productivity differential or are those who have been trained a select group because they appear to be more productive? This is an old question in the literature, which, although clearly not answered within this framework, could be the subject of a separate study. However, the coefficient reflects the importance of job-specific training. Unions have complained that firms have been shortsighted in not training their employees. It is true that the questionnaire responses indicated that only one-third of the workers had received training by the enterprise and that on average the training period was short; almost one-half (47 percent) of the training was for a duration of less than three months.

---

2. See, e.g., Psacharopoulos (1981, 1985) for estimates of the effect of human capital on earnings in the Third World.2

3. The coefficient is practically constant across both regressions (columns 1 and 2 in Table 7.1), indicating that there are no extra rewards in total compensation as compared to base salary.

We also tested for systematic variation in salary of workers in private domestic-owned firms, publicly owned firms, or French, Syro-Lebanese or "other foreign" owned firms by including a variable indicating percent of capital owned by each of these groups. Since these are not dummy variables, the coefficients reflect the percent differential in a worker's earnings brought about by a one percentage increase in, for example, French or Syro-Lebanese ownership of the firm. Consequently, the differentials can be calculated for a firm with ownership by more than one of these categories. Domestic private ownership is used as the base (intercept). The results for base salary and total earnings are presented in Table 7.1.

With respect to base salary, it appears that the publicly owned and Syro-Lebanese firms pay about the same as the private domestic-owned firms, whereas the French and other foreign firms pay significantly higher base salaries--28.4 percent and 30.6 percent more, respectively. However, when calculated with total earnings, the differentials are quite different. Whereas other foreign firms continue to be in the top position (paying 74.8 percent more than private domestic firms), French firms now appear not to be paying more than private domestic firms, and the public firms appear to be paying the second highest average total compensation (66.9 percent more than private domestic firms). We learned from our interviews that the French tend to upgrade employees through the categories (defined in terms of the base salary) faster than the private domestic firms and that they do this instead of awarding more benefits, which would help explain our findings from this regression. Interestingly, and consistent with the information gathered from interviews with managers and union officials, the Syro-Lebanese firms appear to be paying 29.3 percent less than the private domestic firms with respect to total compensation. Therefore, they tend to pay what the government requires them to pay with respect to base salary but add no compensatory supplements or benefits.

We were also interested in assessing the effects of trade variables on workers' earnings, i.e., do firms that export more and/or are more protected from imports pay higher salaries? The results indicate that the firm's rate of effective protection had no significant impact on salary levels. On the other hand, firms that export a higher proportion of their product do tend to pay slightly higher salaries. Firms which export 100 percent of their product pay an average of 21.7 percent more in terms of base salary than those who do not export any of their product, but no more in terms of total compensation.

Finally, we tested for the effect of the firm's profitability and growth on workers' salaries. Profitability and output growth are frequently hypothesized to have a positive effect on wages because they reflect the firm's ability to pay. According to this hypothesis, firms with a greater ability to pay, are able to pay higher wages to otherwise identical workers. However, in testing this hypothesis, one cannot use actual profit data because profit is measured net of actual wages (i.e., it is the revenue minus labor cost minus nonlabor cost). It hence reflects the firm's ability to pay the owners of capital once workers have been paid, rather than capturing the firm's ability to pay workers. In view of this measurement problem, we use the firms net revenue per worker--defined as the revenue minus nonlabor cost per worker--as the appropriate proxy for the firm's ability to pay. The net revenue measure is conceptually more appropriate than profit because it reflects the total "pie" that can be divided between employees and capital owners. It hence defines the maximum amount that could be allocated to wages.

As the estimated coefficients in Table 7.1 indicate, we found no relationship between a firm's net revenue per worker and the base salary (wage) or total compensation. However, firms that have been growing faster do seem to be paying higher base salaries and greater total compensation. The relevant coefficients in Table 7.1 indicate that a 1 percent increase in annual sales growth (over the five years preceding the year of the current observation) would lead to a 0.82 percent increase in base salaries and a 1.02 percent increase in total compensation.

8

Industrial Wages and the Extended Family

The government of Senegal and the trade unions have expended a great deal of effort to protect the wages of industrial workers. While the data presented in Chapter 4 indicate that in recent years this effort has not prevented real wages from falling, questions arise as to the possible effects of this effort and of movements in real wages in general. Since the 1950s, there has been a general belief that a higher income induces larger extended families, such that consumption per capita of the worker and the worker's family remains constant but the consumption of the extended family is slightly increased.

As Pfeffermann (1968) pointed out in his book on industrial workers in Senegal,

> Income figures per worker convey a highly misleading image of standards of living. They merely constitute a first step in an attempt to find out how the standard of living of urban industrial workers compares with that of peasants or other fractions of the population. In a society in which the 'extended family' system prevails, the number of dependents who are fed (and often clothed) at the wage-earners' sole expense, thereby curtailing his income, must be taken into account. (p. 167).

Similarily, since the publication of the influential development models by Todaro (1969) and Harris and Todaro (1970), development economists and policy makers have viewed the relative behavior of urban wages and rural incomes as an important determinant of rural-urban migration. The prediction of these models is that only a fraction of the rural migrants obtain modern sector jobs, while the rest of them live with relatives in the urban area and engage in modern sector job search as well as informal economic activities. An implication of this prediction is that the size of the extended urban family is positively correlated with wages in the urban modern sector, although the impact on welfare (living standards) depends on a number of factors, including earnings in the informal urban sector, rural incomes, etc.

In this chapter we examine the structure of extended families in the Dakar area and analyze the relationship between wages and family size.

8.1 Size of Extended Families

The extended family can be defined as the urban and rural group of persons who partly or entirely depend on the income of one wage earner for subsistence. (Pfeffermann considered only the urban group.) The extended family can be broken down into the nuclear family, comprising the wage earners' wives and children, and the non-nuclear family, comprising other relatives.

An analysis of the size of the extended family alters the image of per capita incomes of individual workers considerably: Both nuclear and non-nuclear families are remarkably large in Senegal --the former as a result of the absence of birth control and the prevalence of polygamy, and the latter because of traditional family solidarity.

8.1.1 Nuclear Families

In examining the number of wives of male workers in our study, we found that the extent of polygamy has not declined during the 23 years since Pfeffermann completed his study; in fact, it may have slightly risen (see Table 8.1).

The typical family in our 1986 sample is very large--possibly larger than the typical family in the 1963 sample of 188 workers from 13 industrial firms. As can be seen in Table 8.2, the average number of children for monogamous workers was 3.08 in 1963 and 5.08 in 1986. For polygamous workers, the average number was nearly twice as large in both years: 5.93 in 1963 and 9.88 in 1986. The total average number of children per worker in Pfeffermann's 1963 sample is 3.41 and in our 1986 sample it is 5.87. Moreover, it appears that fewer Senegalese workers are now childless. Or, conversely, whereas 75 percent of the 1963 sample of workers had one or more children, the figure rose to 91.4 percent in the 1986 sample.

If the 1963 and 1986 samples of workers are representative, they clearly indicate a substantial increase in family size. Not only has the number of children and wives per family increased,

Table 8.1
Senegal: Distribution of Male Workers by Marital Status

	1963	1986
Single	13.8	10.1
One wife	61.2	63.1
Two wives	21.3	20.9
Three or more wives	3.7	4.6
Widowed or divorced	--	1.4

Source: 1963 -- Pfeffermann (1968), p. 168
1986 -- Worker sample, 1986 survey of 17 industrial firms

TABLE 8.2
Senegal: Average Number of Children by Marital Status

	1963	1986
Monogamous	3.08	5.08
Polygamous	5.93	9.88
Total	3.41	5.87

Source: 1963 -- Pfeffermann (1968), p. 168
1986 -- Worker sample, 1986 survey of 17 industrial firms.

but the number of children per worker has also jumped. The rise in the average number of children is consistent with the rapid population growth rates reported by the official statistics (averaging 2.9 percent per year between 1972 and 1982) and it can probably be explained by the constant fertility rate and falling mortality rate.

8.1.2 Non-nuclear Families

The average number of persons maintained at the worker's expense is even greater than the statistics concerning wives and children would suggest. Unemployed relatives enjoy the frequent hospitality of wage earners and often join their households. Non-nuclear families typically consist of elderly relatives and the children of brothers and sisters from the provinces.

The data on the number of people who are regularly fed (and sometimes clothed, in the case of the Pfeffermann data) are astounding by American and European standards. It should be noted that these figures do not include the numerous temporary guests who come to stay in town for periods ranging from a few days to several weeks. Again, if the two samples are representative and, moreover, if the question was indeed asked in the same way, the figures indicate that the size of the extended non-nuclear family has increased since 1963. Pfeffermann found that the average worker in 1963 fed (and sometimes clothed) 9.6 people including himself. In 1986 the average worker fed about 11.9 people including himself (we did not include clothing in our study). Table 8.3 shows the frequency distribution of the workers by size of extended family. Note that 16 percent of the workers each supported 14 or more people in 1963; in 1986 the corresponding proportion rose to 35 percent.

In seeking to explain this pattern, we conjecture that increased size of the extended family is in part a result of the deterioration in Senegal's economic performance. In particular, with rising unemployment, stagnant industrial employment, and a decline in GNP per capita, holding a job has become more of a "privilege." Those who are now in this privileged minority are expected to shoulder heavier family responsibilities. The results of our study are consistent with this explanation and also supports Pfeffermann's finding (1968, p. 221) that there is a pattern between the size of the extended family and the age of the worker. Pfefferman noted that older workers tend to have larger extended families; our data reinforce this relationship. Table 8.4 presents the data from both surveys.

TABLE 8.3

Senegal: Distribution of Workers by Size of Extended Families*

Family Size	1963	1986
2 to 4	14.2	5.2
5 to 7	25.5	12.2
8 to 10	27.5	22.0
11 to 13	17.0	21.2
14 to 16	7.0	15.2
17 or more	9.0	19.8
Total	100.0	100.0

*Extended family is defined as the number of people fed regulary in the worker's home.
Source: 1963 -- Pfeffermann (1968), p. 169.
　　　　1986 -- Worker sample, 1986 survey of 17 industrial firms.

In both samples the proportion of those under 31 who support 10 or more people is far below the proportion of the oldest group (over 40) who support this number. In comparison, only 7 percent of the oldest group were supporting fewer than 6 people in 1986, whereas 43 percent of those under 31 were supporting this relatively small number of people.

How does this pattern--that older workers support larger extended families--help explain the finding (in Table 8.3) that the extended families were considerably larger in 1986 as compared to 1963? The answer can be seen in the last row of Table 8.4, which shows the proportion of workers in each age group for the two survey years. It is clear that workers in 1986 were on average older than workers in 1963. If this age distribution is representative of the entire population of workers in the industrial sector, then a strong presumption exits that the combined forces of government regulation and economic deterioration are halting the employment of new (young) workers

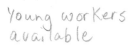
Young workers available

TABLE 8.4

Senegal: Cross Tabulation of Workers' Age and Size of Extended Family

	AGE GROUP					
	1963			1986		
Number of Dependents	Under 31	31 to 40	Over 40	Under 31	31 to 40	Over 40
less than 6	63	22	17	43	15	7
6 to 9	26	35	22	22	29	17
10 and more	11	43	61	35	56	76
Total:	100 %	100 %	100 %	100 %	100 %	100 %
Age Distribution:	29 %	38 %	33 %	20 %	39 %	40 %

Source: 1963 -- Pfeffermann (1968), p. 221.
1986 -- Worker sample, 1986 survey of 17 industrial firms.

and hence creating an older industrial work force in Senegal. The older workers, now a larger proportion of the pool of employed, are bearing a larger share of the family support burden.

However, from Table 8.4 it is also clear that within each age group the burden of responsibility has grown since 1963. For example, whereas only 11 percent of those under 31 supported 10 or more people in 1963, in 1986 over one-third (35 percent) of this group were supporting 10 or more people. Similarly, 76 percent of workers over 40 were supporting 10 or more people in 1986, as compared to just 61 percent for this category in 1963.

8.2 Relationship Between Earnings and the Extended Family

One of the findings from Pfeffermann's (1968) analysis of Senegal's labor market was that some employers believed that higher incomes did not necessarily lead to higher consumption patterns for the worker and the nuclear family. Some employers felt that higher incomes either (1) induced more migration of marginal relatives from the rural areas to the town or (2) induced the worker to increase the number of people supported in the extended family.[1]

After excluding young workers living with their fathers and not in charge of a family, Pfeffermann (1968, p. 219) found that the coefficient of correlation between total wage compensation and the number of people maintained by the worker was 0.319. That is to say for every 1 percent increase in monthly earnings, the typical industrial worker would support 0.3 percent additional people. The relationship existed but was not very strong, as shown

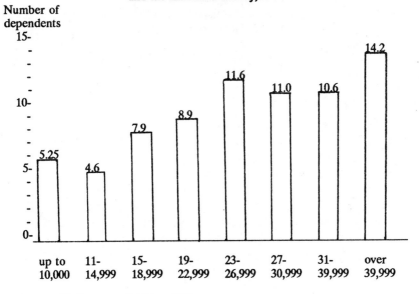

FIGURE 8.1

Senegal: Relationship Between Income and the Extended Family, 1963

Source: Pfeffermann (1968, p. 220)

in Pfeffermann's diagram (reproduced as Figure 8.1) plotting income against number of dependents.

---

1. Pfeffermann claims there is no evidence that the flow of money transferred by industrial workers to relatives in the country increases with income. If anything, the evidence suggests that younger workers tend to send larger amounts of money home than do older ones.

Our data also indicate that the size of the extended family--in terms of number of people fed--increases as income rises, although the relationship is not very strong (see Table 8.5). When we regressed the number of people fed by a Senegalese industrial worker on the worker's base wage and total compensation, we found small but statistically significant coefficients on the wage variables (Table 8.5). However, the wage variables only explain 2 percent of the variance in the number of extended family members. Our data therefore support the hypotheses that industrial wages have a positive effect on the size of the extended family, but the explanatory power of our results is very small.

TABLE 8.5

Senegal: Correlation between Size of Extended Family and Worker's Earnings

(Dependent Variable: Number of people fed)

Independent Variables:	Coefficient	t Statistic
Intercept	11.6	10.6
Total Compensation	.38E-5	2.3
Base Salary	.23E-4	3.4
R^2	.02	
N^*	558	

Source: 1986 -- Worker sample, 1986 survey of 17 industrial firms.
*Sample = Senegalese workers only.

8.3 Secondary Income

The effect of the extended family on the worker's standard of living becomes clearer when one realizes that the earnings a worker receives from a job in the industrial sector usually represent his/her total earnings. In other words, there are relatively few workers who either earn income from a secondary activity or who have a secondary earner in the household. This phenomenon is found in both Pfeffermann's and our data. Pfeffermann (1968) found that 17 percent of his sample occasionally earned some extra money from moonlighting, but in our sample the percentage is much lower--5 percent.[2] Lachaud's (1987) study of workers in state owned firms finds that 11.8 percent of these workers have at least one other source of income. These findings suggest that the extent of moonlighting by industrial workers has been decreasing over time, and that in the 1980s the phenomenon is more prevalent among workers in government owned rather than private firms.

Except in the case of young unmarried workers living in their parents' homes, it is rare to find more than one wage earner per household. As soon as job seekers find stable employment, they set up their own households. In response to our question regarding secondary earners in the household, only 7.7 percent of our sample indicated there were others in the home contributing to the family income. This answer may have been biased downward because the question also asked how much was being contributed by others in the family. However, it is interesting to note that only 12 percent of the women indicated that there was another person in the household contributing to the family income and that 7 percent of the men said their spouse or some other family member was contributing. Pfeffermann found that 2.2 percent of the men in his survey had wives who earned money. As with moonlighting, there appears to be a different pattern of household labor force participation in the case of public sector employees. In particular, Lachaud (1987) finds that about one-third of state enterprise workers in his sample live in households with two or more wage earners and that more than one-half of these secondary earners also work in the public sector.

In summary, neither our nor Pfeffermann's findings strongly support the industrialists' argument that increases in industrial wages do not translate directly into increases in the worker's consumption and standard of living. The cross sectional evidence for 1963 and 1986 indicates at best a weak positive relationship between the size of extended families and income. A simple linear regression of earnings and the size of extended family produces

---

2. This statistic may be another reflection of the deterioration in Senegal's economic conditions.

a very small positive coefficient. Moreover, this regression only explains 2 percent of the variance in extended family size.

What has become very clear from the analysis of the extended family is that its size has increased over time due to the aging of the industrial work force and to the increased number of dependents within each age group. We surmise that the larger extended family "dependency ratio" has arisen from a deterioration in economic conditions in Senegal. The few people who are fortunate to have jobs in the industrial sector are supporting a larger pool of unemployed. Moreover, because of the stagnation in industrial employment, the employed increasingly tend to be older workers.

The extent of moonlighting by industrial workers is limited, as is the frequency of finding more than one income earner in the families of these workers. The exception is the public sector where both phenomenon appear to be more pronounced.

9

Summary and Policy Implications

Senegal's economy has been characterized by long-term stagnation and, in terms of many indicators, even decline. In the industrial labor market, which is the focus of our analysis, the situation has been reflected in the virtually stagnant employment levels of full-time workers and in falling real wages. In view of these developments, the aim of this study has been to describe how Senegal's industrial labor market functions and to examine the effects of the industrial relations system, major institutions, and principal structural factors on economic performance.

9.1 Summary

1. The modern industrial labor market is heavily regulated by the government. The regulations may be aimed at advancing workers' welfare, but it is questionable whether they are having this effect; instead, these regulations have resulted in higher production costs and reduced flexibility of industrial enterprises. The workweek is fixed by law, and managers feel particularly constrained by their inability to lay off workers without first passing through an elaborate and time-consuming process of government approval. The difficulties associated with extending the contracts of temporary (daily) workers are also cited as significant constraints on enterprise behavior.

2. The industrial sector is heavily unionized, but the unions are in many respects weak and workers do not seem to be enthusiastic about their achievements. The major influence of the unions comes from their ability to induce the government to institute and enforce the aforementioned regulations. Their other effect comes from being partners in a highly adversarial system of industrial relations. The system is tripartite (unions-employers-government) in nature and is characterized by deep mistrust--

especially between the unions and employers. Its negative effect on efficiency may appears to be considerable.

3. Low worker effort, motivation, and discipline are cited by managers as major causes for the low labor productivity of Senegal's major industries. This phenomenon is exacerbated by employers' opposition to unions, by the union position that workers must resist employer exploitation, and by the lack of a direct link between performance and remuneration in most enterprises.

4. Another reason for the low level of labor productivity is the limited availability of skilled workers and what appears to be a secular decrease in the provision of on-the-job training by firms. There has been a widening of the total compensation differential between skilled and unskilled employees without a corresponding increase in the skilled labor supply. This low supply response is caused, to a great degree, by limited educational and training opportunities; it can also be attributed to the many difficulties faced by individuals in obtaining credit to finance their training and by the fact that higher income tends to some extent increase the worker's dependency burden through the influx of additional relatives into the extended family.

5. Our econometric estimates provide a number of important insights into the determinants of enterprise productivity:

(i) Overall, one cannot find a significant relationship between total factor productivity (TFP) and enterprise ownership, except in the Syro-Lebanese firms, whose estimated TFP is 30-50 percent below that of other firms; the Syro-Lebanese firms also provide comparatively lower compensation to their employees.

(ii) The impact of ownership on TFP varies by industry: In the food industry, firms with greater Senegalese state and/or foreign (other than French) ownership are more productive than other firms; in the chemical and extraction industries, firms with more French and other foreign ownership are more productive than others; somewhat weaker statistical evidence suggests that firms with "other foreign" ownership tend to be less productive than others in the mechanical industry;

(iii) Effective (trade) protection has a major negative effect on TFP, but enterprise export orientation displays no productivity effect;

(iv) Although non-Africans are on average paid 123-142 percent more than comparable Africans, the estimated marginal

productivity of non-African employees is not significantly different from zero (i.e., removing one non-African employee would not, on average, affect output);
(v) At the industrial-sector level, there was a detectable 5-7 percent average annual decline in TFP during the 1980-1985 period; industry-specific analysis indicates that this decline is evident in the food and mechanical industries, less so in the textile, clothing and leather industry but not at all in the chemical and extraction industries.

6. Because of falling real base wages, the share of the employer-determined wage supplements in total compensation has increased dramatically over time, as did the use of temporary (daily) workers. The latter phenomenon is in part a reaction to the regulatory burden in the labor market. Both phenomena reflect an increase in the decisionmaking flexibility of the employers.

7. The pattern of enterprise ownership has a significant effect on wages (earnings): Non-French, foreign-owned firms pay the highest wages, and local Syro-Lebanese firms pay the lowest; greater export-orientation of a firm is associated with slightly higher base wage but not total earnings; effective protection has no noticeable effect on wages; and finally, there appears to be a highly significant positive relationship between the output growth of the firm and the wages it pays.

8. The size of the industrial worker's extended family has increased over time. Deteriorating economic conditions have increased the "dependency ratio" for each employed worker and through reduced hiring, have created an older labor force that is traditionally expected to support more dependents. Extended families of industrial workers frequently rely on the breadwinners wage as the sole source of income; in the public sector, families are often composed of more than one wage earner.

9.2 Policy Implications

Our qualitative and quantitative findings suggest that Senegal's industry has suffered from a highly adversarial system of industrial and labor relations, excessive government regulations in many areas alongside insufficient government support in others; and numerous misperceptions about the influence of the ethnically diverse labor force and enterprise ownership. Our specific policy conclusions are as follow.

1. The lackluster performance of Senegal's economy and our findings about the industrial labor market underscore the need for effective policies geared toward long-term growth in productivity, employment and earnings. Our numerous discussions with the three social partners give us a definite impression that the country needs to adopt a social compact that would replace the existing highly adversarial system of industrial relations with greater cooperation among the partners. In order to reap the benefits that such a system could bring in terms of greater effort, motivation, and labor productivity, it would be highly desirable to link a part of worker compensation (e.g., the wage supplement) to specific individual or group performance indicators.

2. Because the current regulations contribute to the underutilization of labor, a relaxation of the regulations concerning (paid) hours of work, layoffs, and the use of temporary (daily) workers would, along with transfer payments, improve the welfare of all the parties.

3. The limited extent of training programs for workers is a serious problem that may require government intervention. If firms do not provide optimal amounts of training because they have to bear the training cost and fear being subsequently raided by other firms, some form of collective action or government intervention may be needed.

4. The fact that no particular type of ownership has had a detectable overall productivity advantage, and that certain types of ownership display higher productive efficiency than others in specific industries suggests that policies geared toward changing enterprise ownership or providing preferential treatment to certain types of ownership should be based on a careful examination of specific cases. In particular, case study evidence suggests that a number of enterprises are in need of closure or rehabilitation. However, the econometric evidence indicates that, if productive efficiency is an important criterion, policies in this area cannot be in the form of general rules based solely on ownership (e.g., privatization, Senegalization, preferential treatment of foreign capital, etc.).

5. The econometric findings--that effective protection has a negative effect on enterprise productivity (measured in world prices) and no effect on wages--are strong and precise. Within the

general development policy debate, these findings provide important support for the argument that protection generates inefficiency. They also indicate that this inefficiency is not an economic rent that would be reflected in higher wages. The results suggest that the recent policies aimed at reducing trade protection should have a positive effect on productivity, thus making firms more capable of facing foreign competitors and pass on the benefits of improved efficiency to consumers in the form of lower prices. The overall impact of reduced protection of course depends on a number of factors, including the proportion of firms that survive the adjustment process and the extent of new investment in the country.

6. Our analysis indicates that the extent of an enterprises export orientation is not related to productive efficiency; this finding has important implications for both the general debate about optimal development policy and for Senegal in particular. It contradicts the hypothesis that firms in developing countries can increase their productive efficiency by greater exposure to the world market: On the one hand, because we are able to control for the extent of effective protection in our sample, we are able to test this hypothesis very carefully. On the other hand, because many of the exporting firms in our sample export only to other West African countries, the argument about the efficiency gained from exposure to the world market may not be fully tested.

7. The finding that a large part of Senegal's industrial sector may be experiencing a significant decline in total factor productivity signals a potential long-term problem that should be explored within the ongoing New Industrial Policy. Although it is difficult within the context of this study to pinpoint the exact determinants of this decrease in productivity, our discussions with managers, workers, and trade union leaders suggest that the unwillingness of firms to invest in physical and human capital, in combination with labor market and other governmental regulations, has contributed to the decline.

8. The exodus of the highly paid non-Africans has been welcomed by the proponents of Senegalization and viewed with suspicion by those who perceive the departure as a serious loss of highly productive human capital. Our econometric estimates suggest that, on the margin, the productivity of the non-African labor input is zero. This implies that, although the average contribution of non-Africans may be high, their gradual exodus

appears to have benefited Senegal's economy, with further gains to be realized as the number of non-Africans on the payrolls gradually declines.

9. The dual finding that the return (in terms of earnings) on firm-specific experience is higher than the return on general experience and that the effect of firm-specific training on earnings is high suggests that the New Industrial Policy should give special consideration to retraining the displaced workers within existing firms rather than outside of them.

10. The effect that deteriorating economic conditions is having on the size of workers' extended families suggests that the burdens of rising unemployment and limited hiring of young workers are being borne mostly by the older industrial workers. Economic growth would hence have important benefits for these workers by decreasing the size of their extended families.

11. The finding of a strong positive relationship between output growth and worker earnings suggests that a firm's ability to pay is an important determinant of worker earnings and that the resumption of economic growth would also benefit workers in terms of significant earnings gains.

More generally, the various elements of our study suggest that the careful design of growth-and efficiency-oriented policies, cast in the framework of a broadly acceptable social compact, would significantly benefit both Senegal's economy and its society.

References

Aharoni, Y. (1986) <u>The Evolution and Management of State-Owned Enterprises</u>, Cambridge, MA: Ballinger Publishing Co.

Bloch, P. (1985) "Wage Policy, Wage Structure and Employment in the Public Sector of Senegal," World Bank, CPD Discussion Paper No. 1985-41, May.

Bowden, R., and J. Turkington (1984) <u>Instrumental Variables</u>, Cambridge, U.K.: Cambridge University Press.

Caves, D. W. and L. R. Christensen, (1980): "The Relative Efficiency of public and private Firms in a Competitive Environment: The Case of Canadian Railroads," <u>Journal of Political Economy</u>, vol. 88, pp. 958-976.

Delegation a l'Insertion, a la Reinsertion et a l'Emploi (1988) <u>Incitation des Employeurs et des Travailleurs a la Promotion des Entreprises de la Production et de l'Emploi</u>, Dakar: Presidence de la Republique.

Gersovitz, M. and Waterbury, J. (1987) <u>The Political Economy of Risk and Choice in Senegal</u>, London: Frank Cass Publishing Co.

Harris, J. and Todaro, M. (1970): "Migration, Unemployment and Development: A Two-Sector Analysis," <u>American Economic Review</u>, vol. 60, pp.126-141.

Lachaud, J. P. (1987) "Restructuration des Entreprises Publiques et Ajustements sur le Marche du Travail au Senegal: Des Possibilities a la Mesure des Esperances," Discussion Paper No. 4, International Institute of Labor Studies, Geneva.

Martens, G. G. (1982) "Industrial Relations and the Political Process in Senegal," <u>Research Series No. 70</u>, Geneva: International Institute for Labor Studies.

Mingat, A. (1982) "Rapport Sur les Relations Entre la Formation et l'Emploi au Senegal," Direction de la Statistique, Senegal, November.

Parti Socialiste du Senegal, Conseil National (1984) La Politique de L'Emploi (21 Juillet).

Pfeffermann, Guy (1968) Industrial Labor in the Republic of Senegal, New York: Praeger, Publishers.

Psacharopoulos, G. (1981) "Returns to Education and Updated International Comparison," Comparative Education, vol. 17, no. 3, pp. 321-341.

_____(1985) "Returns to Education: A Furthere International Update and Implications," Journal of Human Resources, vol.20, pp. 583-597.

Svejnar, J. (1984) "The Determinants of Industrial-Sector Earnings in Senegal," Journal of Development Economics, vol. 15, pp. 289-311.

Svejnar, J., and Hariga, M. (1987) "Public vs. Private Ownership, Export Orientation and Enterprise Productivity in a Developing Economy: Evidence from Tunisia," Working Paper No. 217, Department of Economics, University of Pittsburgh.

Todaro, M. (1969) "A Model of Labor migration and Urban Unemployment in less Developed Countries." American Economic Review, vol. 59, no. 1, pp. 138-148.

World Bank (1984) Senegal Country Economic Memorandum.

World Bank (1987) Senegal, An Economy Under Adjustment.

World Bank (1988-1989) World Tables, vol. 1 and 2.

Index

Agriculture, 6
Arbitration, 18

Birth control, 108
Bloch, P., 24, 38
Bonuses. *See* Wages, bonuses

Cameroon, 5
Cap-Vert, 21, 23
CEAO. *See* West African Economic Community
Chemical and extraction industries, 77–82(tables)
Clothing industry. *See* Textile, clothing and leather industries
CNTS. *See* Confederation Nationale des Travailleurs du Senegal
Cobb-Douglas function, 55–56. *See also* Production functions
Confederation Nationale des Travailleurs du Senegal (CNTS), 13, 14, 43, 50–51
Cost of living, 38, 39, 40, 44

Dakar, 21, 108
Directorate of Labor. *See* Ministry of Public Affairs and Labor, Labor Directorate
Droughts, 7, 8

Education, 47, 103–104
EEC. *See* European Economic Community
Employers
  associations of, 14–15
  and income vs. consumption patterns of workers, 112–113, 115
  and wage differentials, 53
  and wage supplements/bonuses, 34
  See also Managers
Employment
  agencies, 15
  contracts, 16–17, 45, 117
  daily, 22, 45, 51, 119
  hiring, 15–16. *See also* Employment, recruitment for
  layoffs, 17, 45, 51
  managerial, 33(table). *See also* Managers; Wages, managers
  national composition of, 31–34, 32(table), 33(table). *See also* Wages, and national groups
  at night, 18
  overtime, 18, 45
  permanent, 16, 25–28, 26–27(tables), 30(table), 31–32, 32–33(table), 34, 35–36(table), 45
  recruitment for, 46–49, 48(table). *See also* Employment, hiring
  regulations. *See* Labor regulations
  seasonal, 22
  sectors, breakdown by, 22
  skills in, 35–36(table), 48(table), 118. *See also* Wages, and skill groups
  in state-owned firms, 49

temporary, 16, 22, 28–30, 28(n), 29–30(tables), 31, 44, 51, 69, 119
total, 22–23
total industrial, 26(table), 32(table)
training for, 34, 47, 104, 118, 120
urban, 24
wages. *See* Wages
workweek, 46, 117
See also Labor force
European Economic Community (EEC), 1, 8
Exports, 9. *See also* Firms, export-oriented; Industrial sector, export-oriented
Extended families
age of workers in, 110, 111–112, 112(table)
definition, 108
and income, 112–116, 113–114(tables)
non-nuclear, 110–112
nuclear, 108–109
number of children in, 108–109, 109(table)
number of wives in, 108, 109(table)
size of, 108–112, 111–112(tables), 114(table), 116, 119
and standard of living, 107, 114–116

Families. *See* Extended families
Firms, modern industrial
definition, 24
expatriate sales of, 24
export-oriented, 55, 57, 66, 70, 105, 121. *See also* Industrial sector, export-oriented
French, 105
number of, 25(table)
ownership, 55, 56, 70, 105, 118, 119, 120
private vs. public, 56, 57
Syro-Lebanese, 56, 57, 105, 118, 119
Fish processing, 8
Food industry, 28(n), 71–76(tables)
French Overseas Labor Code (1952), 13

GDP. *See* Gross Domestic Product
Gross Domestic Product (GDP), 1, 5–8, 6(table)
Groundnuts, 6, 8

Harris, J., 107

Illiteracy, 47
Importateurs et Exportateurs de l'Ouest Africain (SCIMPEX), 13
Imports, 6
Income
and extended family, 113(table)
per capita, 5
See also Wages
Industrial sector
decline in, 11
domestic market, 9, 10
export-oriented, 8, 9, 118. *See also* Firms, modern industrial, export-oriented
informal urban, 1, 22, 24
modern (formal). *See* Modern industrial sector
ownership, 9, 70
production levels, 9, 10(table)
structure and growth, 8–11
traditional, 13, 22
Inflation, 38
IV estimates. *See* Production functions, instrumental variable (IV) estimates
Ivory Coast, 5

Labor Code (1961), 13, 43
contracts, 16–17

hiring regulations, 15–16
layoffs, 17
managers' views of, 45–49
See also Labor regulations
Labor force, 22–24
African, 47
marital status of males in, 109(table)
national composition of, 31–34, 32(table), 33(table)
Senegalese in, 32–33(tables)
urban, 23(table)
See also Employment
Labor Office, 15–16
Labor regulations, 117, 120
and government officials, 43–44
job security, 51
managers' views of, 45–46, 46(table)
reform in, 44
See also Labor Code (1961)
Labor Service. *See* Ministry of Public Affairs and Labor, Labor Service
Labor unions. *See* Trade unions
Lachaud, J. P., 49, 115
Leather industry. *See* Textile, clothing and leather industries

Managers, 33(table), 45–49, 46(table). *See also* Wages, managers
Mechanical industry, 83–88(tables)
Methodology, 25–27, 54, 69
Mingat, Alain, 1, 23
Minimum wage. *See* Wages, *salaire minimum interprofessionel garanti*
Ministry of Finance, 24, 26, 56
Ministry of Planning, 22
Ministry of Public Affairs and Labor, 14
Employment Directorate, 43
Labor Directorate, 17, 38, 43
Labor Service, 45
Modern industrial sector, 1
definition, 2
employment in, 22, 23, 24, 24–34
firms in. *See* Firms, modern industrial
permanent workers in, 25–28, 26(table), 27(table)
vs. traditional sector, 13
See also Industrial sectors
Moonlighting, 115, 116

National Collective Agreement (NCA), 13–14, 18. *See also* Labor Code (1961)
NCA. *See* National Collective Agreement
New Industrial Policy, 44, 121, 122

Oil prices, 8
OLS estimates. *See* Production functions, ordinary least squares (OLS) estimates

Paper and wood industry, 95–100(tables)
Permanent workers. *See* Employment, permanent
Pfeffermann, Guy, 47, 49, 107, 108, 110, 113, 113(n), 115
Phosphate mining, 8
Polygamy, 21, 108, 109(tables)
Population, 7, 21, 109
Production functions, 54, 55
industry specific, 69–70. *See also individual industries*
instrumental variable (IV) estimates, 58–61(table), 67–68(table), 69, 71–73(table), 77–79(table), 83–85(table), 89–91(table), 95–97(table)

ordinary least squares (OLS)
estimates, 62–65(table), 67–68(table), 74–76(table), 80–82(table), 86–88(table), 92–94(table), 98–100(table)
See also Productivity
Productivity
all industries, 58–65(tables), 67–68(table)
non-African, 121
private vs. public firms, 56, 57
and skill groups, 67–68(table)
Syro-Lebanese firms, 56, 57
total factor, 118–119
See also Production functions

Recession, 8
Recruitment. See Employment, recruitment for

SCIMPEX. See Importateurs et Exportateurs de l'Ouest Africain
Sectors
defined, 6(n3)
employment breakdown in, 22
and GDP, 6(table)
industrial. See Industrial sectors
primary, 6, 7, 22–23
public, 7–8, 23
secondary, 5–6, 7, 8, 23. See also Industrial sectors
service, 7
tertiary, 5–6, 23
See also Industrial sectors; Modern industrial sector
Senegal
cost of living in. See Cost of living
demographic data for, 21
economic crisis in, 1
at independence, 5
life expectancy in, 21
liquidity crisis (1980–1981), 8

ministries. See individual ministries
population, 7, 21
productivity. See Productivity
public expenditures, 7–8
rural-urban migration in, 107
trade regime, 9–11
Senegalization, 31, 31(n), 42, 51, 66, 121
Skills. See Employment, skills in; Productivity, and skill groups; Wages, and skill groups
SMIG. See Wages, *salaire minimum interprofessionel garanti*
Social security, 34
Standard of living, 107, 114–116
State Secretariat for Employment, 15–16
Strikes, 14, 18, 49–50
Syndicat des Commercants, 13
SYPAOA. See Syndicat des Patrons de l'Ouest Africain
Syro-Lebanese, 31, 56, 57, 118, 119

Tariffs, 9–10
Temporary workers. See Employment, temporary
Textile, clothing and leather industries, 89–94(tables)
TFP. See Total factor productivity
Todaro, M., 107
Total factor productivity (TFP), 118–119, 121. See also Productivity
Trade protection, 55, 57, 70, 105, 118, 120–121
Trade unions, 14–15, 117
advantages of, 50(table), 103
movement, 13
radical, 43(n)
and training, 104
and wage differentials, 53
workers' views of, 49–50

Training. *See* Employment, training for
Tripartite Commission, 14, 15, 17, 34, 38

Unemployment, 2, 22, 23–24
Union Syndicale des Industriels (UNISYNDI), 13
UNISYNDI. *See* Union Syndicale des Industriels

Wages, 17–19
 African, 101, 101(n)
 base, 14, 17, 34, 36, 37(table), 38–40, 102–103(table), 106, 119
 bonuses, 14, 17–19, 34, 37
 determinants of, 101–106
 and education, 103–104
 and enterprise characteristics, 102–103(table)
 evolution of, 38–42
 and exporting firms, 105–106
 and extended families, 107–116
 French firms, 105
 male-female differential, 101
 managers, 37(table), 39(table), 41, 41–42(table)
 and national groups, 40–42, 41–42(table)
 non-African employees, 40–41, 42, 53, 101, 101(n), 118–119
 and personal characteristics, 102–103(table)
 and productivity, 46, 122
 and profitability and growth of firms, 106
 salaire minimum interprofessionel garanti (SMIG), 14, 17, 38, 44
 Senegalese vs. non-African, 53
 and skill groups, 37(table), 38–42, 39(table), 41–42(table). *See also* Employment, skills in
 supplements, 14, 18–19, 34, 36–37, 37(n), 38
 Syro-Lebanese firms, 57, 105, 119
 temporary vs. permanent wage bills, 29–30, 30(table)
 total compensation, 34, 36–38, 37(table), 39(table), 40–42, 41–42(table), 102–103(table), 105
 and trade protection, 105
West African Economic Community (CEAO), 1, 8
Wood industry. *See* Paper and wood industry
Workers. *See* Labor force
World Bank, 44, 55, 56